Systems & Control: Foundations & Applications

Series Editor

Tamer Başar, University of Illinois at Urbana-Champaign

Editorial Board

R. Srikant

The Mathematics of Internet Congestion Control

Birkhäuser
Boston • Basel • Berlin

R. Srikant
Coordinated Science Laboratory and
  Department of General Engineering
University of Illinois
Urbana, IL 61801
USA

Library of Congress Cataloging-in-Publication Data

Srikant, Rayadurgam.
  The mathematics of Internet congestion control / Rayadurgam Srikant.
    p. cm. – (Systems and control : foundations and applications)
  Includes bibliographical references and index.
    1. Internet–Mathematical models. 2.
  Telecommunication–Traffic–Management–Mathematics. I. Title. II. Systems & control.

  TK5105.875.I57S685   2003
  004.67'8'015118–dc22   2003065233
                                                          CIP

AMS Subject Classifications: 90C25, 93D2, 93A15, 93D30, 93E15, 90B18, 90B10, 90B15

ISBN 0-8176-3227-1   Printed on acid-free paper.

©2004 Birkhäuser Boston   *Birkhäuser*

Printed in the United States of America.   (SB)

9 8 7 6 5 4 3 2 1   SPIN 10925164

Birkhäuser is part of *Springer Science+Business Media*

www.birkhauser.com

To
Amma, Appa
Susie, Katie, and Jenny

Preface

The Transmission Control Protocol (TCP) was introduced in the 1970s to facilitate reliable file transfer in the Internet. However, there was very little in the original TCP to control congestion in the network. If several users started transferring files over a single bottleneck link at a total rate exceeding the capacity of the link, then packets had to be dropped. This resulted in retransmissions of lost packets, which again led to more lost packets. This phenomenon known as *congestion collapse* occurred many times in the mid-1980s, prompting the development of a congestion control mechanism for TCP by Van Jacobson. By all accounts, this has been a remarkably successful algorithm steering the Internet through an era of unprecedented global expansion. However, with access speeds to the Internet having grown by several orders of magnitude over the past decade and round-trip times increasing due to the global nature of the Internet, there is a need to develop a more scalable mechanism for Internet congestion control. Loosely speaking, by scalability, we mean that the protocol should exhibit a provably good behavior which is unaffected by the number of nodes in the Internet, the capacities of the links, and the RTTs (round-trip times) involved. The purpose of this book is to provide an introduction to the significant progress in the mathematical modelling of congestion control which was initiated by work of Frank Kelly in the mid-1990s, and further developed by many researchers since then.

The book draws upon results from three widely-used topics to model the Internet: convex optimization, control theory and probability. It would be difficult to read this book without a level of knowledge equivalent to that of a first undergraduate course in each of these three topics. On the other hand, it does not require much more than a first-level course in these topics to be able to understand most of the material presented in the book. At the end of most chapters, a brief appendix is included with the intent of providing some background material that would be useful in understanding the development in the main body of the book.

While the mathematical modelling of Internet congestion control is a fairly recent topic, due to the importance of the topic, it has attracted a large number of researchers who have made important contributions to this subject. From among this vast body of literature, I have chosen to focus in this book on those approaches which are rooted in the view that congestion control is a mechanism for resource allocation in a network. Even with this framework, there is a large body of work and I have primarily chosen to emphasize scalable, decentralized mechanisms in this book.

Many people have directly or indirectly contributed to my understanding of the topic of mathematical modelling of the Internet. I wish to thank Frank Kelly whose seminal work was the impetus to all the topics discussed in the book. I have learned a lot about congestion control both from reading his papers as well as from interactions with him at conferences and through email discussions. It is a pleasure to acknowledge and thank Tamer Başar for earlier collaboration on the ATM ABR service and more recent collaboration on the primal-dual algorithm for TCP congestion control. I would also like to acknowledge the work of many present and past graduate students who have contributed to many of the results presented in this book. They include Srisankar Kunniyur, Sanjay Shakkottai, Supratim Deb, Damien Polis, Srinivas Shakkottai, Ashvin Lakshmikantha, Julian Shao Liu and Lei Ying. I also gratefully acknowledge the help of Supratim in generating some of the figures used in this book and providing comments on Chapter 8, and Julian for proof-reading the entire manuscript. I would also like to thank Eitan Altman, Carolyn Beck, Geir Dullerud, Peter Key, A.J. Ganesh, Don Towsley and Chris Hollot for collaborations that have contributed to my understanding of the topics in the book. Finally, I thank Bruce Hajek for always being available to discuss any mathematical problem on any topic. My discussions with him have greatly shaped the direction of much of my research in general.

Urbana, IL   *R. Srikant*
October 2003

Contents

1

Introduction

The Internet has evolved from a loose federation of networks used primarily in academic institutions, to a global entity which has revolutionized communication, commerce and computing. Early in the evolution, it was recognized that unrestricted access to the Internet resulted in poor performance in the form of low network utilization and high packet loss rates. This phenomenon known as congestion collapse, led to the development of the first congestion control algorithm for the Internet [39]. The basic idea behind the algorithm was to detect congestion in the network through packet losses. Upon detecting a packet loss, the source reduces its transmission rate; otherwise, it increases the transmission rate. The original algorithm has undergone many minor, but important changes, but the essential features of the algorithm used for the increase and decrease phases of the algorithm have not changed through the various versions of TCP, such as TCP-Tahoe, Reno, NewReno, SACK [17, 54]. An exception to this is the TCP Vegas algorithm which uses queueing delay in the network as the indicator of congestion, instead of packet loss [14]. One of the goals of the book is to understand the dynamics of Jacobson's algorithm through simple mathematical models, and to develop tools and techniques that will improve the algorithm and make it scalable for networks with very large capacities, very large numbers of users, and very large round-trip times.

In parallel with the development of congestion control algorithms for the Internet, congestion control was studied for other data networks of the time. A simple, yet popular, mathematical model for allocating resources in a fair manner between two users sharing a single link was developed by Chiu and Jain in [16]. This was an early algorithm that recognized the connection between congestion control and resource allocation. We present the model and its analysis below.

Consider two sources accessing a link that can serve packets at the rate c packets per sec. Let x_i be the rate at which source i is injecting packets into the network. Suppose that the link provides feedback to the source indicating whether the total arrival rate at the link ($x_1 + x_2$) is greater than the link capacity or not. Thus, the feedback is $I_{x_1+x_2>c}$, the indicator function of the event $x_1 + x_2 > c$. The sources respond to this congestion indication by adjusting their rates as follows: for each i, x_i evolves according to the differential equation

$$\dot{x}_i = I_{x_1+x_2 \leq c} - \beta x_i I_{x_1+x_2>c}. \tag{1.1}$$

To see how this system behaves, define $y = x_1 - x_2$. Then, $y(t)$ evolves according the differential equation

$$\dot{y} = -\beta y I_{x_1+x_2>c}.$$

When $x_1 + x_2 \leq c$, then clearly y does not change with time, thus, $x_1 - x_2$ remains a constant. However, from (1.1), x_1 and x_2 increase. To consider the behavior of x_1 and x_2 when $x_1 + x_2 \geq c$, let $V(y) = y^2$. Then,

$$\dot{V} = -2y\dot{y} = -2y^2 I_{x_1+x_2>c}.$$

Thus, when $x_1 + x_2 > c$, $\dot{V} < 0$ unless $y = 0$. When $x_1 + x_2 \leq 0$, it can be seen from (1.1) that both x_1 and x_2 increase, while $x_1 - x_2$ remains a constant. Thus, with a little bit more work, one can conclude that, as $t \to \infty$, $x_1 + x_2 \to c$, and $y \to 0$. In other words, in steady-state, the link is shared equally between the two sources and the link is fully utilized. The Chiu-Jain algorithm identifies several features of congestion management which are of interest to us:

- Congestion control: The sources control their rates x_1 and x_2 depending on the level of congestion in the network. If the arrival rate at the link is too large, then the sources decrease their transmission rates and if it is too small, then the sources increase their rates.
- Congestion feedback: The network (in this case, a single link) participates in the congestion management process by providing feedback in the form of $I_{x_1+x_2>c}$. Note that the amount of feedback required is minimal; it requires only one bit of information from the link: whether the arrival rate exceeds the capacity at the link or not. The link does not even have to actively participate in the feedback process. If we assume that packets are dropped when the arrival rate exceeds capacity, then the receivers can detect lost packets and inform the source that there is congestion in the network.
- Fairness in resource allocation: The goal of the congestion control algorithm can be viewed as driving the system towards a fair operating point, which in this case corresponds to each user getting half of the available bandwidth.
- Utilization: The link is fully utilized, i.e., at equilibrium, the arrival rate is equal to the available capacity.
- Decentralization: The congestion controllers are decentralized. Each controller needs only one bit of information from the network, but requires no communication with the other controller(s).

The discrete-time version of (1.1) is given by

$$x_i(k+1) = x_i(k) + \delta I_{x_1(k)+x_2(k)\leq c} - \beta\delta I_{x_1(k)+x_2(k)>c},$$

for some small $\delta > 0$. For $c = 1$, $\delta = 0.05$ and $\beta = 1/2$, a plot of the evolution of the user rates is shown in Figure 1.1 starting from the initial condition $x_1 = 0.3$ and $x_2 = 0.1$.

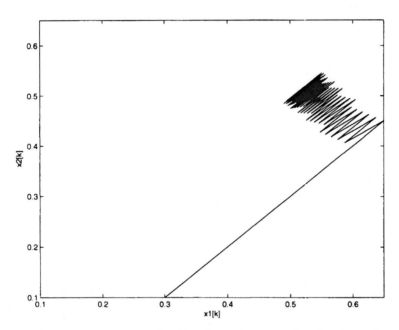

Fig. 1.1. Rate evolution using the Chiu-Jain algorithm. Starting from the point $(0.3, 0.1)$, the system moves towards the point $(0.5, 0.5)$

Next, let us consider the system shown in Figure 1.2. In this system, again there are two sources accessing a single link. However, a packet from source i takes $d_1(i)$ time units to reach the link and it takes $d_2(i)$ time units for the feedback from the link to reach source i. Thus, the dynamics of the source i are given by

$$\dot{x}_i = I_{x(t-d_2(i))\leq c} - \beta x_i I_{x(t-d_2(i))>c},$$

where $x(t)$ is the total arrival rate at the link at time t and is given by

$$x(t) = x_1(t - d_1(1)) + x_2(t - d_1(2)).$$

Following [13], let us consider the evolution of source i's rate at some time $t + d_2(i))$:

$$\dot{x}_i(t + d_2(i)) = I_{x_1(t-d_1(1))+x_2(t-d_1(2))\leq c}$$
$$-\beta x_i(t + d_2(i))I_{x_1(t-d_1(1))+x_2(t-d_1(2))>c},$$

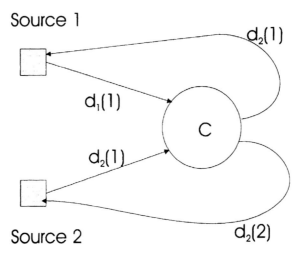

Fig. 1.2. A link with sources having delays in the forward and reverse paths

and define

$$y = x_1(t + d_2(1)) - x_2(t + d_2(2)).$$

If $V(y) = y^2$, then

$$\dot{y} = -2\beta y I_{x_1(t-d_1(1)) + x_2(t-d_1(2)) > c},$$

from which we can conclude as in the delay-free case that $y \to 0$ as $t \to \infty$. In the presence of delay, note that even in steady-state, the rate allocation is not fair at every time instant, but is fair if one compares time-shifted versions of the two rates. In other words, on average, each user gets its fair share of the bandwidth. A discrete version of the algorithm with time step $\delta = 0.005$, $\beta = 0.5$, $d_1(1) = d_2(1) = 2$ discrete time steps, $d_1(2) = d_2(2) = 5$ time steps was simulated and the results are shown in Figure 1.3.

It is difficult to generalize this algorithm and to establish its convergence for more general topology networks. Further, it is not immediately clear if this simple notion of fairness can be generalized to networks with more than one link. However, this algorithm was historically influential in providing the motivation for the congestion control mechanism in [39] which will be studied in detail in a later chapter. The one-bit feedback mechanism proposed by Chiu and Jain and the work reported in [87] are also precursors to the later one-bit feedback schemes such as RED [27].

After the introduction of the congestion control algorithm for TCP, several researchers developed simple mathematical models that led to a better understanding of the dynamics of this algorithm (see, for example, [65, 73, 56]). Building on a resource allocation model by Kelly [48], the seminal paper by Kelly, Mauloo and Tan [52] presented the first mathematical model and analysis of the behavior of congestion control algorithms for general topology networks. Since then, there have been significant developments in the modelling

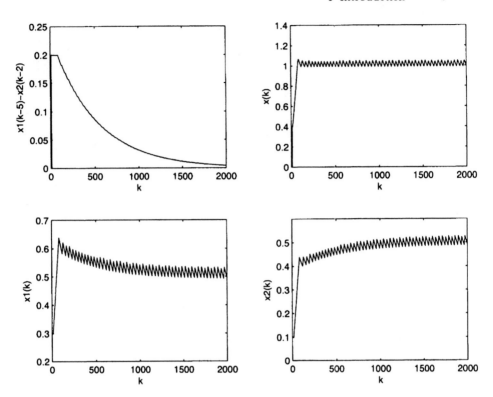

Fig. 1.3. Simulation results of the Chiu-Jain algorithm with feedback delays

of Internet congestion control using tools from operations research and control theory. The goal of this book is to provide a comprehensive introduction to this remarkable progress in the development of an Internet congestion control theory.

2

Resource Allocation

Consider a network where each source r is identified by an origin and a destination between which the user of source r is transferring data. In addition, we suppose that each source r uses a fixed route between its origin and destination, and that the route is specified by a sequence of links. Thus, the index r can be used to denote both the source and the route used by the source. Let x_r be the rate (bits/sec. or bps) at which source r is allowed to transmit data. Each link l in the network has a capacity c_l bps. Given the capacity constraints on the links, the resource allocation problem is one of assigning rates $\{x_r\}$ to the users in a *fair* manner. To illustrate the difficulties in defining a fair resource allocation, we consider the following simple example of a network with three links and two sources.

Example 2.1. Consider the two-link network shown in Figure 2.1. The network

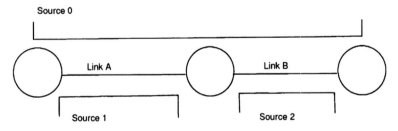

Fig. 2.1. Two-link network

consists of two links A and B and three sources. Source 0's route includes both links A and B, source 1's route consists of only link A and source 2's route consists of link B. Suppose the capacity of link A is $C_A = 2$ and that of link B is $C_B = 1$. Then, a resource allocation that satisfies the link capacity constraints is $x_0 = x_2 = 0.5$ and $x_1 = 1.5$. This resource allocation could be considered fair as explained below.

Suppose that you attempt to divide the capacity of each link among the sources using the link. Then, on link A, sources 0 and 1 would get a rate of 1 each and on link B, sources 0 and 2 would get a rate of 0.5 each. However, since Source 0's route includes both links, it can only transmit at the minimum of the rates that it is assigned on links A and B. Thus, source 0 can only transmit at rate 0.5. Thus, there is still one more unit of capacity remaining to be allocated on link A. This remaining capacity is now allocated to the only other source using that link which is source 1, thus giving $x_0 = 0.5$, $x_1 = 1.5$ and $x_2 = 0.5$. Such a resource allocation is called *max-min fair*. We will precisely define max-min fairness later in this chapter.

While *max-min* fair resource allocation is a particular resource allocation, there are many other ways in which to allocate the available link capacities between the competing sources. For example, $x_0 = 0.25$, $x_1 = 1.75$ and $x_2 = 0.75$. This could be considered fair if, for example, user 2 needs at least 0.75 bps for its application while the added benefit of getting more than 0.25 units is negligible for user 0. Thus, it seems more reasonable to consider the utility of a certain data rate to each user before allocating the network resources. Such a resource allocation will be the subject of the next section.   □

2.1 Resource allocation as an optimization problem

Let us suppose that each user r derives a utility (or benefit) of $U_r(x_r)$ when a data rate x_r is allocated to it. Then it seems reasonable to allocate the network resources to solve the following optimization problem [48]:

$$\max_{\{x_r\} \in \mathcal{S}} \sum_r U_r(x_r) \qquad (2.1)$$

subject to

$$\sum_{r:l \in r} x_r \leq c_l, \qquad l \in \mathcal{L}, \qquad (2.2)$$

$$x_r \geq 0, \qquad r \in \mathcal{S},$$

where $\mathcal{L}$ is the set of all links and $\mathcal{S}$ is the set of all sources in the network. Equation (2.2) simply states that the transmission rates of all the users sharing a link is less than or equal to the capacity of the link. From classical optimization theory [9], the above problem admits a unique solution if $\{U_r(x_r)\}$ are strictly concave functions. Further, it is reasonable to believe that the utility to a user increases when the rate allocated to the user increases. Thus, from now on, we will make the following assumption on the utility functions.

Assumption 2.2 *For each r, $U_r(x_r)$ is a continuously differentiable, non-decreasing, strictly concave function.*   □

We now present examples of various choices for $\{U_r(x_r)\}$ that leads to different resource allocations in a network.

Example 2.3. Proportional fairness. Suppose that $U_r(x_r) = w_r \log x_r$. Let $\{\hat{x}_r\}$ be the optimal solution to the resource allocation problem (2.1) for this set of utility functions, and $\{x_r\}$ be any other set of rates that satisfy the constraints (2.2). Then, from a well-known property of convex functions (see Appendix 2.3), the following is true:

$$\sum_{r \in S} U_r'(\hat{x}_r)(x_r - \hat{x}_r) \le 0.$$

For the choice of utility functions in this example, the above inequality becomes

$$\sum_{r \in S} w_r \frac{x_r - \hat{x}_r}{\hat{x}_r} \le 0.$$

In other words, if we deviate from the optimal allocation $\{\hat{x}_r\}$ to another feasible allocation $\{x_r\}$, then the weighted sum of the proportional changes in each user's rate is less than or equal to zero. Hence, the resource allocation corresponding to $U_r(x_r) = w_r \log x_r$ is called *weighted proportionally fair*. If all w_r's are equal to one, then it is simply called proportionally fair.

Let us now find the proportionally fair resource allocation for the two-link network in Example 2.1. The resource allocation problem is given by

$$\log x_0 + \log x_1 + \log x_2$$

subject to

$$x_0 + x_1 \le 2,$$
$$x_0 + x_2 \le 1,$$

and x_0, x_1 and x_2 are all non-negative. First note that, for this example, it is clear that both constraints would be satisfied with equality. If not, it means that there is unused capacity at one or both of the links and by increasing the rates given to sources 1 and 2, this unused capacity can be made non-zero while increasing the sum of the source utilities. Further since $\log x \to -\infty$ as $x \to 0$, it is clear that the optimal solution will be to allocate non-zero rates to all the users. Thus, we can ignore the non-negativity constraints and use the Lagrange multiplier technique to solve the problem.

Let λ_A and λ_B be the Lagrange multipliers corresponding to the capacity constraints on links A and B, respectively. Then, the Lagrangian for this problem is given by

$$L(\mathbf{x}, \boldsymbol{\lambda}) = \log x_0 + \log x_1 + \log x_2 - \lambda_A(x_0 + x_1) - \lambda_B(x_0 + x_2),$$

where $\mathbf{x}$ is the vector of data rates allocated to the sources and $\boldsymbol{\lambda}$ is the vector of Lagrange multipliers. Now, setting $\frac{\partial L}{\partial x_r} = 0$ for each r gives

$$x_0 = \frac{1}{\lambda_A + \lambda_B}, \quad x_1 = \frac{1}{\lambda_A}, \quad x_2 = \frac{1}{\lambda_B}.$$

Using the fact that $x_0 + x_1 = 2$ and $x_0 + x_2 = 1$ yields

$$\lambda_A = \frac{\sqrt{3}}{\sqrt{3}+1}, \qquad \lambda_B = \sqrt{3}.$$

Thus,

$$\hat{x}_0 = \frac{\sqrt{3}+1}{3+2\sqrt{3}}, \quad \hat{x}_1 = \frac{\sqrt{3}+1}{\sqrt{3}}, \quad \hat{x}_2 = \frac{1}{\sqrt{3}}.$$

□

Example 2.4. Minimum potential delay fairness. Let $U_r(x_r) = -w_r/x_r$. Suppose that source r is attempting to transmit a file of size w_r, then w_r/x_r is the amount of time that it would take to transfer the file if x_r is the rate allocated to it by the network. Thus, the concave program (2.1) can be equivalently thought of as minimizing the sum of file transfer delays of all the sources in the network. Hence, the resulting resource allocation under this class of utility functions is called *minimum potential delay fair.*

Assuming all w_r's are equal to 1, the solution to the resource allocation problem (2.1) for the network in Example 2.1 satisfies

$$\frac{1}{x_0^2} = \lambda_A + \lambda_B,$$

$$\frac{1}{x_1^2} = \lambda_A,$$

$$\frac{1}{x_2^2} = \lambda_B,$$

$$x_0 + x_1 = 2,$$

$$x_0 + x_2 = 1.$$

□

Yet another type of fairness is *max-min fairness* for which we have already presented an example. We formally define it below [40, 10].

Definition 2.5. *A vector of rates $\{x_r\}$ is said to be* max-min fair *if, for any other set of rates $\{y_r\}$ that satisfy the capacity constraints (2.2), the following is true: if $y_s > x_s$ for some $s \in S$, then there exists $p \in S$ such that $x_p \leq x_s$ and $y_p < x_p$.* □

From the above definition, under max-min fair allocation, the only way to increase a source's (say, source s) rate is to decrease the rate allocated to some other source whose rate is less than or equal to source s's rate.

The concept of a bottleneck link for a source is useful in understanding the properties of max-min fair allocation. We define it as follows.

Definition 2.6. *A link $l \in s$ is said to be the bottleneck link for source r if it has the following properties:*

$$\sum_{s:l\in s} x_s = c_l,$$

and

$$x_r \geq x_s, \qquad \forall s \text{ such that } l \in s.$$

□

In other words, link l is a bottleneck link for source r if the link is fully utilized and r has the largest flow rate among all sources using link l.

The following lemma, which is a straightforward consequence of the definition of max-min fairness, states that, under max-min fair allocation, all sources have a bottleneck link.

Lemma 2.7. *A set of rates $\{x_r\}$ is max-min fair if and only if every source r has a bottleneck link.*

Proof. Suppose that $\{x_r\}$ is a max-min fair allocation of rates, but there exists a source r which does not have a bottleneck link. Define

$$y_l = \sum_{s:l\in s} x_s.$$

Therefore, for each $l \in s$, one of the following must be true: either $y_l < c_l$ or there exists $r'(l) \neq r$ such that $x_{r'(l)} > x_r$. For each link l, define the available capacity for source r to be

$$a_l = \begin{cases} c_l - y_l, & y_l < c_l, \\ x_{r'(l)} - x_r, & y_l = c_l. \end{cases}$$

Thus, if x_r is incremented by $\max_{l\in s} a_l$, then only sources whose rates are larger than r will be affected, while all other sources' rates will remain the same. This contradicts the definition of max-min fairness and thus, there must be a bottleneck link for each source.

To prove the lemma in the reverse direction, assume that each source has a bottleneck link under some allocation $\{x_r\}$. Suppose source r is increased, then clearly at least one of the sources at each of its bottleneck links must have their rates reduced to maintain feasibility. But all the sources at a link have rates less than or equal to x_r. Thus, $\{x_r\}$ must be a max-min rate allocation.

□

As we mentioned briefly at the beginning of the chapter, the following algorithm can be used to find the max-min fair rates $\{x_r\}$.

Max-min rate allocation algorithm:

1. Let n_l denote the number of sources in S that use link l. For each l such that $c_l \neq 0$, define the fair share f_l on link l as follows:

$$f_l = \frac{c_l}{n_l}.$$

Thus, the *fair share* of link l is obtained by dividing the link capacity among all the sources equally. Define

$$z_r = \min_{l \in r} f_l.$$

In other words, if in each link l, the rate f_l is allocated to all the sources sharing the link, then r will get z_r. Define z_{min} to be the smallest of these rates in the network, i.e.,

$$z_{min} = \min_r z_r.$$

Let $\tilde{R}$ be the set of sources that have this smallest rate, i.e.,

$$\tilde{R} = \{r \in S : z_r = x_{min}\}.$$

For all $r \in \tilde{R}$, z_r is allocated as the max-min rate, i.e.,

$$x_r = z_r, \qquad r \in \tilde{R}.$$

Set

$$S \leftarrow S \setminus \tilde{R}.$$

If S is empty, stop. Else, continue.
2. For all $l \in \mathcal{L}$, set

$$c_l \leftarrow c_l - \sum_{r \in \tilde{R}, l \in r} x_r.$$

In other words, subtract the capacity used up by the sources that were allocated their max-min rates from the previous step. Note that the links for which $f_l = z_{min}$ will have $c_l = 0$ after this step. Go to Step 1. □

It is easy to see why the resource allocation performed by the above algorithm is max-min fair. We will call the successive execution of Steps 1 and 2, an iteration of the algorithm. The set of sources for which the max-min rates are assigned in Step 1 will be referred to as the *saturated sources*. The set of links with $c_l = 0$ after Step 2 will be called *saturated links*. A little thought shows that the rates allocated to saturated sources in iteration k will be smaller than the rates allocated to saturated sources in iteration $k + 1$. Therefore, during each iteration, the saturated links are the bottleneck links for saturated sources in that iteration. In other words, at every iteration of the algorithm, the saturated sources automatically have bottleneck links due to the nature of the algorithm. Thus, by Lemma 2.7, the allocation resulting from the algorithm is max-min fair.

Let us now apply the max-min rate allocation algorithm to two examples.

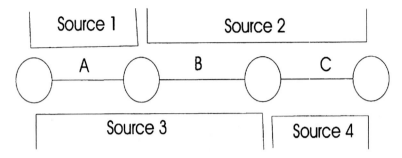

Fig. 2.2. A three-link network with four sources

Example 2.8. Consider the network in Figure 2.1 with $c_A = 10$ and $c_B = 5$. The iterations of the algorithm are shown below:

- Iteration 1 : $S = \{0, 1, 2\}$. $f_A = 5$, and $f_B = 2.5$. Thus, $\tilde{R} = \{0, 2\}$ and $x_0 = x_2 = 2.5$. $c_A \leftarrow 10 - 2.5 = 7.5$ and $c_B \leftarrow 5 - 5 = 0$. Now, $cS \leftarrow S \setminus \tilde{R}\{1\}$.
- Iteration 2 : $f_A = 7.5$. $x_1 = 7.5$ and $\tilde{R} = \{1\}$. Thus, the new $S = \emptyset$ and the algorithm terminates.   □

Example 2.9. Consider the three-link network in Figure 2.2. Let $c_A = 10$, $c_B = 7$, $c_C = 5$. The max-min resource allocation algorithm proceeds as follows:

- $S = \{1, 2, 3, 4\}$, $f_A = 5$, $f_B = 3.5$ and $f_C = 2.5$. Thus, $\tilde{R} = \{2, 4\}$ and $x_2 = x_4 = 2.5$. The new $S = \{1, 3\}$, $c_A = 10$, $c_B = 4.5$ and $c_C = 0$.
- Now, $f_A = 5$ and $f_B = 4.5$. Thus, $\tilde{R} = \{3\}$ and $x_3 = 4.5$. The new $S = \{1\}$, and $c_A = 5.5$ and $c_B = 0$.
- Finally, $f_A = 5.5$ and $x_1 = 5.5$. The new $S = \emptyset$ and the algorithm terminates.   □

2.2 A general class of utility functions

We now introduce a general class of utility functions which subsumes proportional fairness, minimum potential-delay fairness and max-min fairness as special cases [78]. Let the utility function of user r be given by

$$U_r(x_r) = w_r \frac{x^{1-\alpha_r}}{1 - \alpha_r},\tag{2.3}$$

for some $\alpha_r > 0$. Now let us consider some special cases.

Case I: Minimum potential delay fairness, $\alpha_r = 2$, $\forall r$.

In this case, the utility function of a user r is given by

$$U_r(x_r) = -\frac{w_r}{x_r}.$$

Thus, the network's resource allocation is weighted minimum potential delay fair.

Case II: Proportional fairness, $\alpha_r = 1$, $\forall r$.

In this case, the utility function in (2.3) is not well-defined. However, let us consider the derivative of the utility function in the limit as $\alpha_r \to 1$:

$$\lim_{\alpha_r \to 1} U_r'(x_r) = \lim_{\alpha_r \to 1} w_r x_r^{-\alpha_r}$$
$$= \frac{w_r}{x_r}.$$

Thus, in the limit as $\alpha_r \to 1$, the utility function behaves as though $U_r(x_r) = w_r \log x_r$, which leads to weighted proportional fairness. This motivates us to redefine the general class of utility functions (2.3) as follows:

$$U_r(x_r) = \begin{cases} -w_r \dfrac{x^{1-\alpha_r}}{1-\alpha_r}, & \alpha_r > 0, \alpha_r \neq 1, \\ w_r \log x_r, & \alpha_r = 1. \end{cases} \qquad (2.4)$$

Case III: Max-min fairness. $w_r = 1$, $\alpha_r = \alpha$, $\forall r$, $\alpha \to \infty$.

A formal proof of the fact that such a choice of α leads to max-min fairness is given in [78]. We present an informal argument here, along the lines of the proof in [78]. Let $\{x_r(\alpha)\}$ denote the optimal solution to

$$\max_{\{x_r\}} U_r(x_r)$$

subject to

$$\sum_{r:l \in r} x_r \leq C_l, \qquad \forall l,$$

and

$$x_r \geq 0, \forall r.$$

Let $\{y_r\}$ be a feasible allocation of rates, i.e., one where all the link capacity constraints are satisfied. Further, suppose that $\mathbf{y} \neq \mathbf{x}(\alpha)$ and that there exists an s such that $y_s \geq x_s(\alpha)$ for all sufficiently large α. Since $\{x_r(\alpha)\}$ is optimal, we have

$$\sum_r \frac{y_r - x_r(\alpha)}{x_r^\alpha(\alpha)} < 0.$$

Thus,

$$\frac{y_s - x_s(\alpha)}{x_s^\alpha(\alpha)} < -\sum_{r \neq s} \frac{y_r - x_r(\alpha)}{x_r^\alpha(\alpha)}.$$

Since, we have assumed that $y_s - x_s(\alpha) > 0$, we can rewrite the above inequality as

$$1 < \sum_{r \neq s} -\left(\frac{y_r - x_r(\alpha)}{y_s - x_s(\alpha)}\right)\left(\frac{x_s(\alpha)}{x_r(\alpha)}\right)^\alpha$$

$$\leq \sum_{\substack{r \neq s \\ \text{only positive terms}}} -\left(\frac{y_r - x_r(\alpha)}{y_s - x_s(\alpha)}\right)\left(\frac{x_s(\alpha)}{x_r(\alpha)}\right)^\alpha.$$

For the above inequality to hold, there must exist a p such that

$$x_p(\infty) \leq x_s(\infty).$$

Otherwise, all the terms on the right-hand side will go to zero, whereas the left-hand side will be 1. Further, since we are only considering strictly positive terms in the right-hand side, if

$$y_s > x_s(\infty),$$

then y_p must be strictly less than $x_p(\infty)$. In other words, if $y_s > x_s(\infty)$, there exists $p \in S$ such that

$$y_p < x_p(\infty) \leq x_s(\infty).$$

From the definition of max-min fairness in Definition 2.5, the set of rates $\{x_r(\infty)\}$ is max-min fair.

Consider the network in Figure 2.1 with $C_A = 10$, and $C_B = 5$. The max-min rate allocation for this network is given by $x_0 = 2.5$, $x_1 = 7.5$, and $x_2 = 2.5$. Suppose instead we allocate rates according to the solution to

$$\max_{\{x_r\}} \frac{x_1^{1-\alpha}}{1-\alpha} + \frac{x_2^{1-\alpha}}{1-\alpha} + \frac{x_3^{1-\alpha}}{1-\alpha},$$

subject to

$$x_0 + x_1 \leq 10, \tag{2.5}$$

$$x_0 + x_2 \leq 5, \tag{2.6}$$

$$x_1, x_2, x_3 \geq 0. \tag{2.7}$$

The Lagrangian is given by

$$L(x_0, x_1, x_2, p_A, p_B) = \frac{x_1^{1-\alpha}}{1-\alpha} + \frac{x_2^{1-\alpha}}{1-\alpha} + \frac{x_3^{1-\alpha}}{1-\alpha}$$
$$- p_A(x_0 + x_1 - 10) - p_B(x_0 + x_2 - 5).$$

For this problem, it is easy to see that the link capacity constraints will be satisfied with equality. Thus, the solution to the optimization problem is the positive solution to the following set of equations:

$$\frac{\partial L}{\partial x_0} = 0 \Rightarrow \frac{1}{x_0^\alpha} = p_A + p_B,$$

$$\frac{\partial L}{\partial x_1} = 0 \Rightarrow \frac{1}{x_1^\alpha} = p_A,$$

$$\frac{\partial L}{\partial x_2} = 0 \Rightarrow \frac{1}{x_2^\alpha} = p_B.$$

along with (2.5)-(2.6). Table 2.2 presents the solution to this set of equations for different values of α. From the table, it can be seen that the rate allocation

α	x_0	x_1	x_2	p_A	p_B
1	2.11	7.89	2.89	0.13	0.35
2	2.43	7.57	2.57	1.75×10^{-2}	0.15
5	2.50	7.50	2.50	4.21×10^{-5}	1.02×10^{-2}

Table 2.1. Table showing max-min fair allocation in the limit as $\alpha \to \infty$.

indeed converges to the max-min fair rate allocation as $\alpha \to \infty$. The Lagrange multipliers have the following interpretation: in the limit as $\alpha \to \infty$, $(1/p_A)^{1/\alpha}$ is the fair share on link A and $(1/p_B)^{1/\alpha}$ is the fair share on link B. At $\alpha = 5$,

$$\left(\frac{1}{p_A}\right)^{1/\alpha} = 7.5011 \text{ and } \left(\frac{1}{p_B}\right)^{1/\alpha} = 2.5010,$$

which are very close to the max-min fair shares on the two links that were obtained in the previous section.

So far, we have seen how different resource allocation schemes can be achieved with different choices of utility functions. If a network has access to the utility functions of all the users and knows capacities of all the links in the network, then resource allocation can be performed by solving the optimization problems presented in this chapter. However, this is clearly unreasonable since this requires centralized knowledge of all the parameters of the network. A more reasonable set of assumptions would be the following:

- Each user knows its own utility function.
- Each router knows the total arrival rate on its links.
- There is a protocol that allows the network to convey some information about the resource usage on a user's route to the user.

In the next chapter, we will present several algorithms which operate under these assumptions and still solve the resource allocation problem precisely. As we will see, the key idea in all of these algorithms is that each link computes the

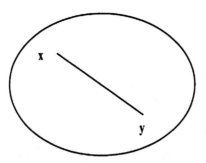

Fig. 2.3. An example of a convex set

Lagrange multipliers corresponding to its capacity constraint and the sources use the knowledge of the sum of the Lagrange multipliers on their routes to compute the optimal transmission rates.

2.3 Appendix: Convex optimization

In this appendix, we define convex sets and convex functions, and briefly summarize some of their properties which would be useful to follow the material in this book. For more details, the readers are referred to [9].

Definition 2.10. *A set $C \subset \Re^n$ is said to be convex if the following property holds:*

$$\alpha\mathbf{x} + (1-\alpha)\mathbf{y} \in C, \qquad \forall \mathbf{x}, \mathbf{y} \in C, \alpha \in [0,1].$$

□

A convex set can be visualized as follows: suppose you consider two points in the set and draw a straight line between the two points; then all points on the straight line also lie within the set. This is illustrated in Figure 2.3.

Definition 2.11. *Consider a function $f : C \to \Re$, where $C \in \Re^n$ is a convex set. The function is said to be convex if*

$$f(\alpha\mathbf{x} + (1-\alpha)\mathbf{y}) \le \alpha f(\mathbf{x}) + (1-\alpha)f(\mathbf{y}), \qquad \forall \mathbf{x}, \mathbf{y} \in C, \quad \forall \alpha \in [0,1].$$

The function is said to be strictly convex if the above inequality is strict when $\mathbf{x} \neq \mathbf{y}$ and $\alpha \in (0,1)$. □

Definition 2.12. *Consider a function $f : C \to \Re$, where $C \in \Re^n$ is a convex set. The function is said to be concave if $-f$ is convex, or equivalently,*

$$f(\alpha\mathbf{x} + (1-\alpha)\mathbf{y}) \ge \alpha f(\mathbf{x}) + (1-\alpha)f(\mathbf{y}), \qquad \forall \mathbf{x}, \mathbf{y} \in C.$$

The function is said to be strictly concave if $-f$ is strictly convex, or equivalently, if the above inequality is strict when $\mathbf{x} \neq \mathbf{y}$ and $\alpha \in (0,1)$. □

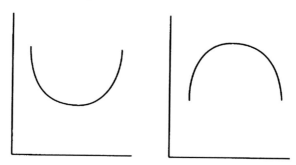

Fig. 2.4. Example of convex and concave functions. The function of the left is convex and the function on the right is concave

Convex and concave functions can be visualized as follows. A strictly convex function looks like a bowl. In other words, if any two points on the graph of a convex function are connected by a straight line, then the line lies above all the points in between the two points. On the other hand, a strictly concave function looks like an inverted bowl. In other words, if any two points on the graph of a strictly concave function are connected by a straight line, then the line lies above all the points in between the two points. Examples of convex and concave functions are shown in Figure 2.4.

Differentiable convex functions have two properties that we will use in this book. We will state these properties in the following lemmas without proofs.

Lemma 2.13. *A function* $f : C \to \Re$, *where* $C \subset \Re^n$, *is convex if and only if the following holds:*

$$f(\mathbf{x}) + (\mathbf{y} - \mathbf{x})^T \nabla f(\mathbf{x}) \leq f(\mathbf{y}), \qquad \forall \mathbf{x}, \mathbf{y} \in C,$$

where

$$\nabla f(\mathbf{x}) = \left[\frac{\partial f}{\partial x_1}, \frac{\partial f}{\partial x_2}, \ldots, \frac{\partial f}{\partial x_n} \right]'.$$

The function is strictly convex if the above inequality is strict when $\mathbf{x} \neq \mathbf{y}$. □

The above property states that the first-order Taylor's series expansion provides a lower bound on the value of a convex function. Recall that if f is concave, then $-f$ is convex. Thus, the inequality in the above lemma will be reversed if f is a concave function.

A special case of Lemma 2.13 occurs when $\mathbf{y} = \hat{\mathbf{x}}$, where $\hat{\mathbf{x}}$ is a point at which $f(x)$ achieves its global minimum over C. In this case, since $f(\hat{\mathbf{x}}) \leq f(\mathbf{x})$ $\forall \mathbf{x} \in C$, we have

$$(\hat{\mathbf{x}} - \mathbf{x})^T \nabla f(\mathbf{x}) \leq 0.$$

For a concave function $f(x)$ which achieves its maximum at $\hat{\mathbf{x}}$, the property becomes

$$(\hat{\mathbf{x}} - \mathbf{x})^T \nabla f(\mathbf{x}) \geq 0. \tag{2.8}$$

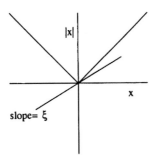

Fig. 2.5. Example of a sub-gradient

For convex functions that are not differentiable, it may still be possible to define a vector that "acts" like its gradient vector. Such a vector is called a *sub-gradient*. A $\boldsymbol{\xi}$ is called a sub-gradient of a convex function f defined over a convex set C if

$$f(\mathbf{x}) + (\mathbf{y} - \mathbf{x})^T \boldsymbol{\xi} \leq f(\mathbf{y}), \qquad \forall \mathbf{x}, \mathbf{y} \in C. \tag{2.9}$$

Note that the sub-gradient plays the role of the gradient in Lemma 2.13. For a fixed $\mathbf{x}$, the equation

$$\mathbf{z} = f(\mathbf{x}) + (\mathbf{y} - \mathbf{x})^T \boldsymbol{\xi}$$

viewed as a function from $\mathbf{y}$ to $\mathbf{z}$ defines a line with slope (gradient) $\boldsymbol{\xi}$ passing through the point $\mathbf{x}$. The definition of sub-gradient simply states that the graph of f lies above this line. There may be more than one sub-gradient at a point $\mathbf{x}$. However, if $f(\mathbf{x})$ is differentiable at $\mathbf{x}$, then the sub-gradient is unique and is equal to the gradient $\nabla f(\mathbf{x})$.

The widely-used function to illustrate the concept of a sub-gradient is $|x|$. From Figure 2.5, any number in the range $[-1, 1]$ is a sub-gradient at the point $x = 0$. Any line with such a slope touches $|x|$ at $x = 0$ and the graph of $|x|$ lies above such a line. However, for values of $x \neq 0$, the gradient is well defined. Thus, for $x < 0$, the unique sub-gradient (which is also the gradient) is -1 and for $x > 0$, the unique sub-gradient/gradient is $+1$.

Now, we will turn our attention to the maximization of concave functions over a convex set. The following lemma will be useful to us.

Lemma 2.14. *Consider a concave function $f(x)$ defined over a convex set $C \subset \Re^n$. The following statements are true:*

- *If $\hat{\mathbf{x}}$ is a local maximum of $f(\mathbf{x})$ over C, then it is also a global maximum over C. If $f(\mathbf{x})$ is strictly concave, then $\hat{\mathbf{x}}$ is also the unique global maximum over C.*
- *If C is compact (closed and bounded), then a global maximum exists over C.*

□

A particular type of maximization problem that we will encounter throughout this book is one where the objective function is strictly concave, while the constraints are linear. In the rest of this appendix, we will consider the following such problem:

$$\max_{\mathbf{x}} f(\mathbf{x}) \tag{2.10}$$

subject to

$$R\mathbf{x} \leq \mathbf{c},$$
$$H\mathbf{x} = 0.$$

where $f(\mathbf{x})$ is a differentiable concave function and R and H are matrices of appropriate dimensions. It is easy to verify that the feasible set, i.e., the values of $\mathbf{x}$ which satisfy the constraints, is convex. The first step in solving such a problem is to form the Lagrangian function L given by

$$L(\mathbf{x}, \boldsymbol{\lambda}, \boldsymbol{\mu}) = f(\mathbf{x}) - \boldsymbol{\lambda}^T (R\mathbf{x} - \mathbf{c}) - \boldsymbol{\mu}^T H\mathbf{x}, \tag{2.11}$$

where $\boldsymbol{\lambda} \geq 0$ is called the Lagrange multiplier. The Karush-Kuhn-Tucker theorem states that, if $\mathbf{x}$ is a solution to (2.10), then it has to satisfy the following conditions:

$$\begin{aligned}
\frac{\partial L}{\partial \mathbf{x}} = 0, &\Rightarrow \nabla f(\mathbf{x}) - R^T \boldsymbol{\lambda} - H^T \boldsymbol{\mu} = 0, \\
&\boldsymbol{\lambda}^T (R\mathbf{x} - \mathbf{c}) = 0, \\
&R\mathbf{x} \leq \mathbf{c}, \\
&H\mathbf{x} = 0, \\
&\boldsymbol{\lambda} \geq 0.
\end{aligned} \tag{2.12}$$

Since the objective function is concave and the feasible set is convex, the above conditions are also sufficient conditions for optimality. In other words, if $\mathbf{x}$ satisfies (2.12), then it solves (2.10).

The final topic of the appendix is duality. Consider the function

$$D(\boldsymbol{\lambda}, \boldsymbol{\mu}) = \max_{\mathbf{x} \in C} L(\mathbf{x}, \boldsymbol{\lambda}, \boldsymbol{\mu}), \tag{2.13}$$

where C is the set of all $\mathbf{x}$ that satisfy the constraints in (2.10). This is called the dual function. We note that

$$\inf_{\boldsymbol{\lambda} \geq 0, \boldsymbol{\mu}} D(\boldsymbol{\lambda}, \boldsymbol{\mu}) \geq f(\hat{\mathbf{x}}),$$

where $\hat{\mathbf{x}}$ is the solution to (2.10). To see this, note that, if $\mathbf{x}$ satisfies the constraints in (2.10), then

$$f(\mathbf{x}) \leq f(\mathbf{x}) - \boldsymbol{\lambda}^T (R\mathbf{x} - \mathbf{c}) - \boldsymbol{\mu}^T H\mathbf{x}.$$

Thus,

$$\max_{\mathbf{x} \in C} f(\mathbf{x}) \leq \max_{\mathbf{x} \in C} \left[f(\mathbf{x}) - \boldsymbol{\lambda}^T (R\mathbf{x} - \mathbf{c}) - \boldsymbol{\mu}^T H\mathbf{x} \right],$$

and further

$$\max_{\mathbf{x} \in C} f(\mathbf{x}) \leq \inf_{\boldsymbol{\lambda} \geq 0, \boldsymbol{\mu}} \max_{\mathbf{x} \in C} \left[f(\mathbf{x}) - \boldsymbol{\lambda}^T (R\mathbf{x} - \mathbf{c}) - \boldsymbol{\mu}^T H\mathbf{x} \right] = \inf_{\boldsymbol{\lambda} \geq 0, \boldsymbol{\mu}} D(\boldsymbol{\lambda}, \boldsymbol{\mu}).$$

This result is called the *weak duality* theorem. Under certain conditions called the *Slater constraint qualification* conditions, we further have

$$\inf_{\boldsymbol{\lambda} \geq 0, \boldsymbol{\mu}} D(\boldsymbol{\lambda}, \boldsymbol{\mu}) = f(\hat{\mathbf{x}}),$$

i.e., the optimal primal and dual objectives are equal. The Slater constraint qualification conditions are always satisfied for a problem with a concave objective function and linear constraints. Thus, the optimal primal and dual objectives are equal for the problem in (2.10).

3

Congestion Control: A decentralized solution to the resource allocation problem

For the development in this section, it is convenient to represent the capacity constraints (2.2) in vector form. To this end, let R be a matrix that is defined as follows: the $(r, l)^{\text{th}}$ entry of R is 1 if source r's route passes through link l and is zero otherwise. Thus, R is an $|\mathcal{S}| \times |\mathcal{L}|$ *routing matrix* which provides information on the links contained in all the source's routes in the network. Let y_l be the total arrival rate of traffic at link l. Then, the vector of link rates $\mathbf{y}$ is given by the relationship

$$\mathbf{y} = R\mathbf{x}.$$

3.1 Primal algorithm

Instead of solving the resource allocation problem (2.1) exactly, let us consider the following problem:

$$V(\mathbf{x}) = \sum_{r \in \mathcal{S}} U_r(x_r) - \sum_{l \in \mathcal{L}} \int_0^{\sum_{s:l \in s} x_s} f_l(y) dy, \tag{3.1}$$

where we interpret $f_l(\cdot)$ as the *penalty function or the barrier function* [9] as the case may be, or simply as the *price* for sending traffic at rate $\sum_{s:l \in s} x_s$ on link l. We make the following assumption on the functions $\{f_l(\cdot)\}$.

Assumption 3.1 *For each $l \in \mathcal{L}$, $f_l(\cdot)$ is a non-decreasing, continuous function such that*

$$\int_0^y f_l(x) dx \to \infty \qquad \text{as } y \to \infty.$$

□

The above assumption is intuitive: it simply states that as the load on the link increases, the penalty on the link does not decrease and further that the penalty should be non-zero for large rates. Now we can state the following lemma.

Lemma 3.2. *Under Assumptions 2.2 and 3.1, the function $V(\mathbf{x})$ defined in (3.1) is strictly concave.*

Proof. We recall from Lemma 2.13 that a differentiable function $g(\mathbf{x})$ defined over a convex set $\mathcal{C} \subset \Re^n$ is convex if and only if the following holds:

$$g(\mathbf{z}) + \sum_{i=1}^{n}(x_i - z_i)\frac{\partial g}{\partial y_i}(\mathbf{z}) \leq g(\mathbf{x}) \qquad \forall \mathbf{x}, \mathbf{z} \in \Re^n.$$

Consider

$$g(\mathbf{x}) = \sum_{l \in \mathcal{L}} \int_0^{\sum_{s:l \in s} x_s} f_l(y)dy.$$

Since $\{f_l(\cdot)\}$ are non-decreasing functions,

$$g(\mathbf{x}) - g(\mathbf{z}) = \sum_{l \in \mathcal{L}} \int_0^{\sum_{s:l \in s} x_s} f_l(y)dy - \int_0^{\sum_{s:l \in s} z_s} f_l(y)dy$$

$$= \sum_{l \in \mathcal{L}} \int_{\sum_{s:l \in s} z_s}^{\sum_{s:l \in s} x_s} f_l(y)dy$$

$$\geq \sum_{l \in \mathcal{L}} f_l\left(\sum_{s:l \in s} z_s\right)\left(\sum_{s:l \in s} x_s - \sum_{s:l \in s} z_s\right) \qquad (3.2)$$

$$= \sum_{l \in \mathcal{L}}\sum_{s:l \in s} f_l\left(\sum_{s:l \in s} z_s\right)(x_s - z_s)$$

$$= \sum_{r \in \mathcal{S}}\sum_{l:l \in r} f_l\left(\sum_{s:l \in s} z_s\right)(x_r - z_r) \qquad (3.3)$$

$$= \sum_{r \in \mathcal{S}}(x_r - z_r)\frac{\partial g}{\partial z_r}(\mathbf{z}). \qquad (3.4)$$

In the above sequence of expressions, (3.2) follows from the fact that f_l's are non-decreasing functions and (3.3) follows from the previous line by interchanging the order of the two summations. Thus, we have verified that $g(\cdot)$ is a convex function, or equivalently, $-g(\cdot)$ is a concave function. Since, by our assumption, $U_r(\cdot)$ is strictly concave, $V(\cdot)$ is the sum of a strictly concave function and a concave function. Thus, $V(\cdot)$ is strictly concave. $\qquad \square$

For the rest of this book, unless otherwise stated, we will also make the following assumption.

Assumption 3.3 *For all r, $U_r(x_r) \to -\infty$ as $x_r \to 0$.*

Consider the following problem:

$$\max_{\{x_r \geq 0\}} V(\mathbf{x}). \qquad (3.5)$$

By Lemma 3.2, $V(\mathbf{x})$ is strictly concave. Further $V(\mathbf{x}) \to -\infty$ as $\|x\| \to 0$, since we have assumed that $U_r(x_r) \to -\infty$ as $x_r \to 0$. Also, $V(\mathbf{x}) \to \infty$ as $\|x\| \to \infty$, due to our assumption on the price functions f_l. Thus, (3.5) has a unique solution that lies in the interior of the set $\mathbf{x} \geq 0$. The optimal source rates satisfy

$$\frac{\partial V}{\partial x_r} = 0, \qquad r \in \mathcal{S}.$$

This gives the following set of equations:

$$U_r'(x_r) - \sum_{l:l \in r} f_l \left(\sum_{s:l \in s} x_s \right) = 0, \qquad r \in \mathcal{S}. \tag{3.6}$$

Clearly, the above set of equations is difficult to solve in a centralized manner. It requires knowledge of all the sources' utility functions, their routes and the link penalty functions f_l's. Next, we will see how congestion controllers compute the solution to (3.5) in a decentralized manner.

Let us suppose that each router (node) in the network has the ability to monitor the traffic on each of its links and compute y_l for each link l connected to it. Let $p_l(t)$ denote the price for using link l at time t. In other words,

$$p_l(t) = f_l(y_l(t)).$$

Define the price of a route to be the sum of the prices of the links on the route. Now, suppose that the sources and the network have a protocol that allows each source to obtain the sum of the link prices along its route. One way to convey the route prices would be to have a field in each packet's header which can be changed by the routers to convey the price information. For instance, the source may set the price field equal to zero when it transmits a packet. Each router on the packet's route could then add the link prices to this field so that, when the packet reaches the destination, the field would contain the route price. The destination could then simply transmit this information back to the source in the acknowledgment (ack) packet. The drawback with this protocol is that the field would have to consist of a large number of bits to convey the price accurately. However, it can be shown that the price can be conveyed using just a single bit of information while incurring a larger variance in the source rates. We will postpone the discussion of the one-bit price communication protocol to a later section in this chapter.

Let q_r denote the route price on route r, i.e.,

$$q_r = \sum_{l:l \in r} p_l,$$

or, in vector notation, the vector of route prices is given

$$\mathbf{q} = R^T \mathbf{p},$$

where R is the routing matrix defined earlier. Using this notation, the condition (3.6) can be rewritten as

$$U'_r(x_r) - q_r = 0, \qquad r \in \mathcal{S}. \tag{3.7}$$

Consider the following congestion control algorithm for solving the set of equations (3.7): for each $r \in \mathcal{S}$,

$$\dot{x}_r = k_r(x_r)\left(U'_r(x_r) - q_r(t)\right), \tag{3.8}$$

where $k_r(x)$ is any non-decreasing, continuous function such that $k_r(x) > 0$ for any $x_r > 0$. Note that the right-hand side of the differential equation (3.8) describing the congestion control algorithm is simply the partial derivative of the objective function (3.1) with respect to x_r, multiplied by a scaling function $k_r(x_r)$. We note the congestion controller is the well-known gradient algorithm that is widely used in optimization theory [9].

The following theorem shows that the congestion control algorithm is globally asymptotically stable, i.e., starting from any initial condition, in the limit as $t \to \infty$, the set of sources $\{x_r(t)\}$ will converge to the set of non-zero rates $\{hx_r\}$ that maximize the objective $V(\mathbf{x})$ in (3.1).

Theorem 3.4. *Under Assumptions 2.2 and 3.1, starting from any initial condition $\{x_r(0) \geq 0\}$, the congestion control algorithm (3.8) will converge to the unique solution of (3.5), i.e., $\mathbf{x}(t) \to \hat{\mathbf{x}}$ as $t \to \infty$.*

Proof. Recall from Lemma 3.2, $V(\mathbf{x})$ is a strictly concave function and thus, (3.5) admits a unique solution. From Assumption 3.3, the solution satisfies $\hat{x}_r > 0 \; \forall r$. Further,

$$\dot{V} = \sum_{r \in \mathcal{S}} \frac{\partial V}{\partial x_r} \dot{x}_r = \sum_{r \in \mathcal{S}} k_r(x_r)\left(U'_r(x_r) - q_r\right)^2 \geq 0.$$

Further, note that $\dot{V} > 0$ for $\mathbf{x} \neq \hat{\mathbf{x}}$ and is equal to zero for $\mathbf{x} = \hat{\mathbf{x}}$. In other words, V is an increasing function which has a unique maximum and once it reaches its maximum, it stays there. Thus, we can reasonably expect $V(\mathbf{x})$ to reach $V(\hat{\mathbf{x}})$ eventually. Since V has a unique maximum, this automatically would imply that $x(t) \to \hat{\mathbf{x}}$ as $t \to \infty$. More precisely, it is easy to verify that the conditions of Lyapunov's theorem (given in the Appendix in Theorem 3.15) are satisfied and thus, the trajectories of (3.8) converge to $\hat{\mathbf{x}}$, starting from any initial condition $\mathbf{x}(0) \geq 0$. $\qquad\square$

In the penalty function method, the link prices $\{p_l\}$ are computed based on price functions $\{f_l(x_l)\}$, and thus, while it solves the problem (3.5), it does not solve the original resource allocation problem (2.1) formulated in the previous section exactly. In practice, this may not be an important issue: the network models are approximations of phenomena that occur in real networks and therefore, a precise solution to (2.1) may not be necessary. However, the algorithms that compute the solution to (2.1) have useful interpretations in the Internet and therefore, we present them in the subsequent sections in this chapter. The practical interpretations of these algorithms will be deferred to a later chapter.

3.2 Dual algorithm

From the Karush-Kuhn-Tucker conditions, the solution to (2.1) satisfies the following equations:

$$x_r = U_r'^{-1}(q_r), \tag{3.9}$$

and

$$p_l(y_l - c_l) = 0, \tag{3.10}$$

and $p_l \geq 0$, where the relationship between $\mathbf{x}$ and $\mathbf{y}$ and $\mathbf{p}$ and $\mathbf{q}$ is given by $\mathbf{y} = R\mathbf{x}$ and $\mathbf{q} = R^T\mathbf{p}$. To solve for $\{p_l\}$ and $\{x_r\}$ which solve this set of equations, we use the following algorithm at each link, along with (3.9) at each source:

$$\dot{p}_l = h_l(p_l)(y_l - c_l)_{p_l}^+, \tag{3.11}$$

where $(g(x))_x^+$ is used to denote the following:

$$(g(x))_x^+ = \begin{cases} g(x), & x > 0, \\ \max(g(x), 0), & x = 0, \end{cases}$$

and $h_l(p_l) > 0$ is a non-decreasing continuous function. Again, this solution is in a decentralized form: the source algorithm (3.9) requires only the knowledge of the price on its route and the link algorithm (3.11) needs only the total arrival rate at its link to compute the price. Note that the link price algorithm (3.11) has an intuitive interpretation: when the link arrival rate is greater than the link capacity, the link price increases. On the other hand, when the link arrival rate is less than the link capacity, the link price decreases. The reason that (3.9) and (3.11) is called the dual algorithm is as follows. Let us consider the dual of the resource allocation problem (2.1) and (2.2). From Appendix 2.3, the dual is given by

$$D(\mathbf{p}) = \max_{\{x_r\}} \sum_r U_r(x_r) - \sum_l p_l \left(\sum_{s:l \in s} x_s - c_l \right)$$

$$= \max_{\{x_r\}} \sum_r \left(U_r(x_r) - x_r \sum_{l:l \in r} p_l \right) + \sum_l p_l c_l.$$

Thus,

$$D(\mathbf{p}) = \sum_r B_r(q_r) + \sum_l p_l c_l \tag{3.12}$$

where

$$B_r(q_r) = \max_{x_r \geq 0} U_r(x_r) - x_r q_r. \tag{3.13}$$

Thus, the dual problem is given by

$$\min_{\mathbf{p} \geq 0} D(\mathbf{p}).$$

To explicitly obtain an expression for $B_r(q_r)$, note that the solution to the maximization problem in (3.13) is given by

$$U_r'(x_r) = q_r.$$

Thus,

$$B_r(q_r) = U_r\left(U_r'^{-1}(q_r)\right) - U_r'^{-1}(q_r)q_r.$$

Next, let us compute the derivative of $U_r'^{-1}(q_r)$ with respect to q_r. Suppose

$$x_r = U_r'^{-1}(q_r),$$

then

$$q_r = U_r'(x_r).$$

Therefore,

$$1 = U_r''(x_r)\frac{dx_r}{dq_r},$$

or equivalently,

$$\frac{dx_r}{dq_r} = \frac{1}{U_r''(U_r'^{-1}(q_r))}.$$

Next, let us consider

$$\begin{aligned}
\frac{dB_r}{dq_r} &= \frac{U_r'(U_r'^{-1}(q_r))}{U_r''(q_r)} - \frac{q_r}{U_r''(x_r)} - U_r'^{-1}(q_r) \\
&= U_r'^{-1}(q_r) \\
&= x_r.
\end{aligned}$$

Thus,

$$\frac{\partial D}{\partial p_l} = c_l - y_l.$$

Now, (3.9) and (3.11) can be interpreted as a gradient descent algorithm to solve (3.12).

The dual algorithm presented here is a special case of the dual algorithm originally presented in [52], see also [50]. A discrete-time version of the algorithm was proved to be stable in [69, 45] and in [99] for the case of the logarithmic utilities. The following theorem proves the convergence of the dual algorithm, following the proof in [82].

Theorem 3.5. *Suppose that the routing matrix R has full row rank. In other words, given $\mathbf{q}$, there exists a unique $\mathbf{p}$ such that $\mathbf{q} = R^T\mathbf{p}$. Under this assumption, the dual algorithm is globally asymptotically stable.*

Proof. We will use "hats" to denote the optimal values. For example $\{\hat{x}_r\}$ is the set of optimal rates which solve (2.1). We have already argued that $\{\mathbf{x}_r\}$ is unique. Since $\mathbf{q}_r = U_r(\mathbf{x}_r)$, $\mathbf{q}$ is also unique. Further, since $\hat{\mathbf{q}} = R^T\hat{\mathbf{p}}$, under the full row rank assumption on R, $\{\hat{p}_l\}$ is also unique. From the KKT (Karush-Kuhn-Tucker) conditions, this implies that there exists a unique $\{\hat{x}_r\}$ and $\{\hat{p}_l\}$ that satisfy the following condition: at each link l, either

$$\hat{y}_l = c_l$$

or

$$\hat{y}_l < c_l \quad \text{and} \quad \hat{p}_l = 0.$$

Now, consider the Lyapunov function

$$V = \sum_{l \in \mathcal{L}}(c_l - \hat{y}_l)p_l + \sum_{r \in \mathcal{S}}\int_{\hat{q}_r}^{q_r}(\hat{x}_r - (U_r')^{-1}(\sigma))d\sigma.$$

Thus,

$$\begin{aligned}
\frac{dV}{dt} &= \sum_l(c_l - \hat{y}_l)\dot{p}_l + \sum_r(\hat{x}_r - (U_r')^{-1}(q_r))\dot{q}_r \\
&= (\mathbf{c} - \hat{\mathbf{y}})^T\dot{\mathbf{p}} + (\hat{\mathbf{x}} - \mathbf{x})^T\dot{\mathbf{q}} \\
&= (\mathbf{c} - \hat{\mathbf{y}})^T\dot{\mathbf{p}} + (\hat{\mathbf{x}} - \mathbf{x})^T R^T\dot{\mathbf{p}} \\
&= (\mathbf{c} - \hat{\mathbf{y}})^T\dot{\mathbf{p}} + (\hat{\mathbf{y}} - \mathbf{y})^T\dot{\mathbf{p}} \\
&= (\mathbf{c} - \mathbf{y})^T\dot{\mathbf{p}} \\
&= (\mathbf{c} - \mathbf{y})^T H(p)(\mathbf{y} - \mathbf{c})_\mathbf{p}^+ \\
&\leq 0,
\end{aligned}$$

where $H(p) = diag\{h_l(p_l)\}$. Further, $\dot{V} = 0$ only when each link satisfies the condition $y_l = c_l$ or $y_l < c_l$ and $p_l = 0$. This is the optimality condition for the resource allocation problem. Thus, the system converges to the unique optimal solution of the resource allocation problem. □

3.3 Exact penalty functions

The dual solution solves the resource allocation problem (2.1)-(2.2) exactly, whereas the penalty function formulation solves it only approximately. The question then arises as to whether the penalty functions $\{f_l\}$ can be chosen such that the solution to the penalty function formulation of the resource allocation problem can solve the original resource allocation problem exactly.

Suppose that, for each link l, we let the price p_l be a function of both the total arrival rate at the link as well as another parameter $\tilde{c}_l$, i.e., $p_l = f_l(y_l, \tilde{c}_l)$, where f_l is an increasing function of y_l and a decreasing function of a non-negative parameter $\tilde{c}_l$. An example of such a function is

$$f_l(y_l, \tilde{c}_l) = \left(\frac{y_l}{\tilde{c}_l}\right)^{B_l}.$$

When $y_l < \tilde{c}_l$, the above expression gives the steady-state overflow probability in an $M/M/1$ queue with arrival rate y_l, service rate $\tilde{c}_l$ and buffer threshold B_l. Due to this interpretation, we often refer to the parameter $\tilde{c}_l$ as the *virtual link capacity*, or simply *virtual capacity*, as opposed to the true link capacity c_l.

In this section, we will show that the parameters $\{\tilde{c}_l\}$ can be *adaptively* chosen such that the penalty function solution solves the resource allocation problem exactly. Towards this end, consider the following adaptation for $\tilde{c}_l$:

$$\dot{q} = \alpha_l \, (c_l - y_l)^+_{\tilde{c}_l}, \tag{3.14}$$

where $\alpha_l > 0$ is a step-size parameter. The rationale behind this adaptive scheme is as follows: if the link arrival rate is greater than the link capacity, then the link price must be increased to signal the sources to reduce their transmission rate. This is done by decreasing $\tilde{c}_l$ since f_l is a decreasing function of $\tilde{c}_l$. On the other hand, when the arrival rate at a link is less than the link capacity, the price is decreased by increasing the virtual capacity $\tilde{c}_l$. Due to the interpretation of the $\tilde{c}_l$ as the virtual capacity of link l, the adaptive algorithm (3.14) is called the *adaptive virtual queue* or AVQ algorithm.

The following theorem establishes the stability of (3.8) and (3.14) [96]. An alternative proof of a modified version of the AVQ algorithm was provided earlier in [61].

Theorem 3.6. *If the routing matrix R has full row rank, the source rates generated by (3.8) and (3.14) converge to the solution of (2.1)-(2.2), starting from any initial state.*

Proof. The resource allocation problem (2.1) has a unique solution. From the Karush-Kuhn-Tucker conditions, we have

$$U'(\hat{x}_r) = \hat{q}_r,$$

and thus, $\hat{q}_r$ is unique for each r. If the routing matrix R has full row rank, then the Lagrange multiplier $\hat{p}_l$ for each link l is unique. To avoid unnecessary technicalities, we will assume that

$$f_l(\hat{y}_l, \hat{\tilde{c}}_l) = \hat{p}_l$$

has a unique solution $\hat{\tilde{c}}_l < \infty$ for each l. Now, consider the Lyapunov function

$$V(\mathbf{x}, \tilde{\mathbf{c}}) = V_1(\mathbf{x}, \tilde{\mathbf{c}}) + V_2(\mathbf{x}, \tilde{\mathbf{c}}),$$

where

$$V_1 = \sum_{r \in \mathcal{S}} \int_{\hat{x}_r}^{x_r} \frac{U_r'(\sigma)}{k_r(\sigma)} (\sigma - \hat{x}_r) d\sigma,$$

and

$$V_2 = \sum_l \frac{1}{\alpha_l} \int_{\hat{\tilde{c}}_l}^{\tilde{c}_l} (f_l(\hat{y}_l, \hat{\tilde{c}}_l) - f_l(\hat{y}_l, z)) dz.$$

We first observe that

$$\begin{aligned}
\frac{dV_1}{dt} &= \sum_r (U_r'(x_r) - q_r)(x_r - \hat{x}_r) \\
&= \sum_r (U_r'(x_r) - U_r'(\hat{x}_r))(x_r - \hat{x}_r) + \sum_r (\hat{q}_r - q_r)(x_r - \hat{x}_r) \\
&= \sum_r (U_r'(x_r) - U_r'(\hat{x}_r))(x_r - \hat{x}_r) + (\hat{\mathbf{q}} - \mathbf{q})^T (\mathbf{x} - \hat{\mathbf{x}}) \\
&= \sum_r (U_r'(x_r) - U_r'(\hat{x}_r))(x_r - \hat{x}_r) + (\hat{\mathbf{p}} - \mathbf{p})^T (\mathbf{y} - \hat{\mathbf{y}}).
\end{aligned}$$

Next, we consider V_2 and note that

$$\begin{aligned}
\frac{dV_2}{dt} &= \sum_l (\hat{p}_l - f_l(\hat{y}_l, \tilde{c}_l))(c_l - y_l)_{\tilde{c}_l}^+ \\
&\overset{(a)}{\leq} \sum_l (\hat{p}_l - f_l(\hat{y}_l, \tilde{c}_l))(c_l - y_l) \\
&= \sum_l (\hat{p}_l - f_l(\hat{y}_l, \tilde{c}_l))(\hat{y}_l - y_l) + \sum_l (\hat{p}_l - f_l(\hat{y}_l, \tilde{c}_l))(c_l - \hat{y}_l) \\
&\overset{(b)}{\leq} \sum_l (\hat{p}_l - f_l(\hat{y}_l, \tilde{c}_l))(\hat{y}_l - y_l) \\
&\leq \sum_l (\hat{p}_l - p_l)(\hat{y}_l - y_l) + \underbrace{(p_l - f_l(\hat{y}_l, \tilde{c}_l))(\hat{y}_l - y_l)}_{\leq 0, \text{ since } f_l \uparrow \text{ as } y_l \uparrow} \\
&\leq \sum_l (\hat{p}_l - p_l)(\hat{y}_l - y_l) \\
&= (\mathbf{p} - \hat{\mathbf{p}})^T (\mathbf{y} - \hat{\mathbf{y}}).
\end{aligned}$$

In the above set of inequalities, (a) is derived from the following reasoning: when the projection (i.e., the restriction that $c_l - y_l$ should be non-negative when $\tilde{c}_l = 0$) is inactive, clearly (a) is valid. When the projection is active, $\tilde{c}_l = 0$, which implies that $f_l(\hat{y}_l, \tilde{c}_l)$ is ∞ and thus,

$$(\hat{p}_l - f_l(\hat{y}_l, \tilde{c}_l))(c_l - y_l)_{\tilde{c}_l}^+ = 0$$

while

$$(\hat{p}_l - f_l(\hat{y}_l, \tilde{c}_l))(\hat{y}_l - y_l) \geq 0.$$

To understand (b), note that, when $\hat{y}_l < c_l$, $\hat{p}_l$ must be zero, thus making

$$(\hat{p}_l - f_l(\hat{y}_l, \tilde{c}_l))(c_l - \hat{y}_l) \leq 0.$$

Therefore,

$$
\begin{aligned}
\frac{dV}{dt} &= \frac{dV_1}{dt} + \frac{dV_2}{dt} \\
&\leq (U'_r(x_r) - U'_r(\hat{x}_r))(x_r - \hat{x}_r) \\
&\leq 0,
\end{aligned}
$$

since $U'_r(\cdot)$ is a decreasing function. Further, $\frac{dV}{dt} = 0$ only when $x_r = \hat{x}_r$, $\forall r$. Thus, stability follows from Lyapunov's theorem. □

3.4 Primal-dual approach

Another approach for solving the resource allocation problem exactly is to use the primal algorithm at the sources and the dual algorithm at the links. Thus, we continue to use the same algorithm at the source as in the primal penalty function formulation:

$$\dot{x}_r = k_r(x_r)\left(U'_r(x_r) - q_r\right),\tag{3.15}$$

but the link prices are generated according to the dual algorithm's link law:

$$\dot{p}_l = h_l(p_l)\left(y_l - c_l\right)^+_{p_l}.\tag{3.16}$$

Theorem 3.7. *The primal-dual algorithm is globally, asymptotically stable.*

Proof. Consider the candidate Lyapunov function

$$V(\mathbf{x}, \mathbf{p}) = \sum_{r \in \mathcal{S}} \int_{\hat{x}_r}^{x_r} \frac{1}{k_r(\sigma)}(\sigma - \hat{x}_r)d\sigma + \sum_{l \in \mathcal{L}} \int_{\hat{p}_l}^{p_l} \frac{1}{h_l(\beta)}(\beta - \hat{p}_l)d\beta.$$

Thus,

$$
\begin{aligned}
\frac{dV}{dt} &= \sum_r (U'_r(x_r) - q_r)(x_r - \hat{x}_r) + \sum_l (p_l - \hat{p}_l)(y_l - c_l)^+_{p_l} \\
&\leq \sum_r (U'_r(x_r) - q_r)(x_r - \hat{x}_r) + \sum_l (p_l - \hat{p}_l)(y_l - c_l) \\
&= (\mathbf{q} - \hat{\mathbf{q}})^T(\mathbf{x} - \hat{\mathbf{x}}) + (\mathbf{p} - \hat{\mathbf{p}})^T(\mathbf{y} - \hat{\mathbf{y}}) \\
&\quad + \underbrace{\sum_r (U'_r(x_r) - \hat{q}_r)(x_r - \hat{x}_r)}_{\leq 0,\ \text{since } U'_r \downarrow \text{ as } x_r \uparrow} + \underbrace{\sum_l (p_l - \hat{p}_l)(\hat{y}_l - c_l)}_{\leq 0,\ \text{since } \hat{p}_l = 0 \text{ if } \hat{y}_l < c_l} \\
&\leq 0.
\end{aligned}
$$

Further, $\frac{dV}{dt} = 0$ only when $x_r = \hat{x}_r$ and for each link l, either $p_l = \hat{p}_l$ or $\hat{y}_l = c_l$. □

The primal-dual approach to Internet congestion control was introduced in [96, 2], and was generalized in [68].

3.5 Other variations in the primal approach

3.5.1 Exponentially averaged rate feedback

In Section 3.1, we assumed that the link price p_l is a function of the instantaneous link arrival rate y_l, i.e., $p_l = f_l(y_l)$. However, one could also imagine a scenario where the link computes its price based on a weighted average of the arrival rate. In other words, $p_l = f_l(z_l)$, where z_l is a weighted average of the arrival rate computed using the following equation:

$$\varepsilon_l \dot{z}_l = -z_l + y_l, \tag{3.17}$$

where ε_l is a *weight* parameter. To see that this is indeed an averaging equation, consider the discretized version of the previous equation, where we assume for convenience that the discretization is over unit time intervals. Thus, we obtain

$$z(t+1) = \left(1 - \frac{1}{\varepsilon_l}\right) z_l + \frac{1}{\varepsilon_l} y_l.$$

Thus, the averaging equation (3.17) can be thought of as giving a weight of $1 - 1/\varepsilon$ to the previous estimate of the link arrival rate and a weight of $1/\varepsilon$ to the current rate. The following theorem shows that the system of differential equations (3.8)-(3.17), with $p_l = f_l(z_l)$, is globally, asymptotically stable.

Theorem 3.8. *Assume that $f_l(z_l)$ is strictly increasing at $z_l = \hat{z}_l$ for all l. The congestion controller (3.8) where the link prices $\{p_l\}$ are determined as functions of the averaged rate $\{z_l\}$, given by $p_l = f_l(z_l)$, where z_l has the dynamics given in (3.17) is globally, asymptotically stable.*

Proof. Consider the candidate Lyapunov function

$$V(\mathbf{x}) = \sum_r \int_{\hat{x}_r}^{x_r} \frac{1}{k_r(x_r)} (\sigma - \hat{x}_r) d\sigma + \sum_l \varepsilon_l \int_{\hat{z}_l}^{z_l} (f_l(\beta) - f_l(\hat{z}_l)) d\beta.$$

Next, note that

$$\frac{dV}{dt} = \sum_r (x_r - \hat{x}_r)(U_r'(x_r) - q_r) + \sum_l (f_l(z_l) - f_l(\hat{z}_l))(-z_l + y_l)$$

$$= \sum_r (x_r - \hat{x}_r)(U_r'(x_r) - \hat{q}_r) + \sum_r (\hat{q}_r - q_r)(x_r - \hat{x}_r)$$

$$+ \sum_l (p_l - \hat{p}_l)(-z_l + y_l)$$

$$\leq (\hat{\mathbf{q}} - \mathbf{q})^T (\mathbf{x} - \hat{\mathbf{x}}) + (\mathbf{p} - \hat{\mathbf{p}})^T(-\mathbf{z} + \hat{\mathbf{y}}) + (\mathbf{p} - \hat{\mathbf{p}})^T(-\hat{\mathbf{y}} + \mathbf{y})$$

$$= (\mathbf{p} - \hat{\mathbf{p}})^T (\hat{\mathbf{y}} - \mathbf{z})$$

$$= \sum_l (f_l(z_l) - f_l(\hat{z}_l))(\hat{z}_l - z_l)$$

$$\leq 0,$$

where the last line follows from the fact that f_l is an increasing function and the previous line follows from the fact that $\hat{z}_l = \hat{y}_l$. Since $f_l(z_l)$ is assumed to be strictly increasing at $\hat{z}_l$, $\frac{dV}{dt} = 0$ only when $z_l = \hat{z}_l \ \forall l$. This does not necessarily prove that the source rates converge to $\{\hat{x}_r\}$. However, a generalization of Lyapunov's theorem, called the LaSalle invariance principle [55], states that the system converges to the largest invariant set that is contained in the set of all points where $\frac{dV}{dt} = 0$. An invariant set is one such that, if the state of the system lies in the set at $t = 0$, then the system remains in the set for all $t \geq 0$. It is easy to verify that the only invariant point is given by $x_r = \hat{x}_r$, for all r. □

3.5.2 One-bit marking in the primal method

In the penalty function method, we assumed that the link price of a path is the sum of the link prices of the links in the path, and that this information can be conveyed to the source using some protocol. However, today's Internet protocols allow only one bit of information per packet to convey link prices back to the source. To deal with this situation, it is convenient to assume that the link prices are real numbers in the interval $[0, 1]$. Thus, the link price p_l can be thought of as the probability with which a link marks a packet. *Marking* refers to the action of the link by which it flips a bit in the packet header from a 0 to a 1 to indicate congestion. Instead of the sum of the link prices, let q_r denote the probability with which a packet is marked on route r. Thus,

$$q_r = 1 - \prod_{l:l \in r}(1 - p_l).$$

Further, suppose that a source increases its transmission rate only when it receives an unmarked packet. Then, the source law becomes

$$\dot{x}_r = k_r(x_r)\left((1 - q_r)U_r'(x_r) - q_r\right). \tag{3.18}$$

The following theorem establishes the stability of this congestion controller.

Theorem 3.9. *The set of congestion controllers (3.18) is globally, asymptotically stable.*

Proof. It is convenient to rewrite (3.18) as

$$\dot{x}_r = k_r(x_r)(1 + U'_r(x_r)) \left(\prod_{l \in r}(1 - p_l) - \frac{1}{1 + U'_r(x_r)} \right).$$

Now, consider the Lyapunov function

$$V(\mathbf{x}) = -\sum_r \int_0^{x_r} \log(1 + U'_r(\sigma))d\sigma - \sum_l \int_0^{y_l} \log(1 - f_l(\beta))d\beta. \quad (3.19)$$

Since $U'_r(\cdot)$ and $-f_l(\cdot)$ are decreasing functions, in a manner similar to the proof of Lemma 3.2, it is easy to show that $V(\mathbf{x})$ is a convex function. Now,

$$\frac{dV}{dt} = \sum_r \kappa_r(x_r)(1 + U'_r(x_r))$$

$$\times \left(\log \frac{1}{1 + U'_r(x_r)} - \log \prod_{l:l \in r}(1 - p_l) \right) \left(-\frac{1}{1 + U'_r(x_r)} + \prod_{l:l \in r}(1 - p_l) \right).$$

Since $\log(\cdot)$ is an increasing function,

$$(\log u - \log v)(u - v) \geq 0, \qquad u, v \in \Re.$$

Thus, $dV/dt \leq 0$ and is zero only at the equilibrium state of (3.18). $\quad\square$

Note that the congestion controllers given by (3.18) do not solve the penalty function formulation (3.1), but rather converge to the minimum of the convex function given in (3.19).

3.6 REM: A one-bit marking scheme

In the previous section, we showed the convergence of the primal congestion controller when the link prices belong to $[0, 1]$ and can thus be interpreted as probabilities. On the other hand, suppose that the link prices are general non-negative numbers, not necessarily less than or equal to 1. A simple scheme was developed for this case in [5] to convey the path price to a source using just one bit of information. We present this scheme, called Random Early Marking (REM), in this section.

Consider a route $r \in S$. To be able to implement any of the congestion controllers that we have seen in this chapter, the source has to know the path price q_r given by

$$q_r = \sum_{l:l \in r} p_l.$$

Thus, the problem is to convey q_r to source r using just one bit per packet. Suppose that each link l marks each packet that passes through it with probability $1 - e^{-\gamma p_l}$, where $\gamma > 0$ is some fixed parameter. Thus, the probability of marking a packet increases as the link price p_l increases. A packet that traverses all the links in route r is not marked only if it is not marked on any link on its path. Thus, the probability that a packet is not marked on route r is given by

$$\prod_{l:l\in r} e^{-\gamma p_l} = e^{-\gamma \sum_{l:l\in r} p_l} = e^{-\gamma q_r}.$$

Thus, by measuring the fraction of packets that are not marked, each source r can infer the price of its route.

3.7 Multipath routing

In this section, we consider networks where multiple paths are available for each user between its source and destination, and the user can direct its flow along these paths using source routing. The amount of flow on each path is determined by the user in response to congestion indications from the routers on the path. Currently, source routing is not supported in most routers in the Internet and so we have to *overlay* the network with routers that allow source routing.

We refer to a data transfer between two nodes as a *flow*. Each flow may use multiple *paths* to transfer data. Obviously many flows can exist between a node-pair, but it is not necessary that all of them share the same path-set. An example of this would be when some users subscribe to the multipath service but others do not.

Let us consider a network where there could be multiple routes between each source-destination pair. Our goal is to find an implementable, decentralized control for such a network with multipath routing. Unlike before, we now explicitly distinguish between a user and the routes used by a user. We use the notation x_r to denote the flow on route r, and $s(r)$ to denote the source of route r. As in [52], we wish to maximize the total system utility given by

$$\sum_i U_i \left(\sum_{i:s(r)=i} x_r \right) \qquad (3.20)$$

subject to

$$\sum_{r:l\in r} x_r \leq c_l, \qquad \forall l,$$

and $x_r \geq 0$ $\forall r$. Even if we assume that the U_i's are strictly concave functions, the objective in (3.20) is not strictly concave in $\mathbf{x}$. Thus, there exists a possibility of multiple optimal solutions to the problem (3.20).

The penalty function form of the problem (3.20) is given by

$$\mathcal{U}(\mathbf{x}) = \sum_i U_i \left(\sum_{r:s(r)=i} x_r \right) - \sum_l \int_0^{\sum_{r:l\in r} x_r} p_l(y)\, dy. \qquad (3.21)$$

In the above formulation, the utility function of user i depends on the sum of the rates on all its routes. Notice that $\mathcal{U}(\mathbf{x})$ may not be a strictly concave function of $\mathbf{x}$ even if the U_i's are a strictly concave function. Thus, there may be multiple values of $\mathbf{x}$ that maximize (3.21).

If there is a unique maximum to (3.21), as before, the following congestion control algorithm can be proved to be stable and it can be shown that its equilibrium point maximizes the above net system utility function:

$$\dot{x}_r = \kappa_r \left(1 - \frac{1}{U'_{s(r)}\left(\sum_{\beta:s(\beta)=s(r)} x_\beta \right)} \sum_{l\in r} p_l \left(\sum_{\alpha:l\in\alpha} x_\alpha \right) \right). \qquad (3.22)$$

Multiple solutions to the maximization of (3.21) can also be handled, but that is a little bit more complicated.

3.8 Multirate multicast congestion control

In the previous sections, we only studied the congestion control problem for unicast sources, i.e., sources which transmit from an origin node to a single destination node. However, there are many applications, where the data from a single node has to be transmitted to many destinations simultaneously. An example would be a popular music concert that may be viewed by many people in the world. There are two ways for the network to perform this point-to-multipoint transmission:

- Unicast method: One can view the transmission from one source to say, n endpoints, as n separate unicast transmissions. In this case, the congestion control is no different from the problems considered in the previous sections.
- Multicast: The reason that unicast transmission is not desirable for point-to-multipoint transmission is that this results in wastage of network resources. Consider the Y-network shown in Figure 3.1. Suppose we need to transmit a packet from node A to receivers at nodes C and D. Then, under unicast transmission, two copies of the packet are made at node A and each packet is transmitted from node A, one to node C and the other to node D. Thus, the link AB must process two packets. On the other hand, under multicast transmission, a single packet is transmitted from node A and when the packet reaches node B, two copies of the packet are created and one is sent to node C and the other is sent to D. Thus, each link only has to transmit one packet under multicast transmission. In general, if the source A transmits at the rate of x packets, then under

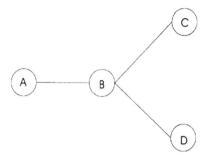

Fig. 3.1. A multicast tree with a source at node A, a junction node at node B, and receivers at nodes C and D

unicast transmission, the load on link AB would be $2x$, whereas, under multicast transmission, the load on link AB would only be x.

Once the decision has been made to transmit using multicast, there are still two further methods of multicast transmission that one can use to transmit, based on the application:

- Single-rate multicast: In this form of multicast, the intended application is online sharing of continuously-modified documents among a large group of people. For such applications, the data are useful only if all of the data are available at all the receivers. Further, the data should reach the receivers at nearly the same time. Thus, if x is the transmission rate at node A in Figure 3.1, then the received rates at nodes C and D are also x.

- Multirate multicast: This form of multicast is used for real-time transmission, where each receiver can receive data at different rates. For example, suppose that link BC is a 56 kbps modem connection and link BD is a broadband home connection which receives at rate 1 Mbps, and the transmission from the source at node A is real-time video. Then, it would be appropriate to have a mechanism by which the transmitter transmits at a single rate, while the receivers receive at different rates with the high-speed receiver being able to receive a better-quality video than a slow-speed receiver. Such a mechanism is described in [74], where the video is transmitted in "layers." Packets in the lowest layer are minimally essential to be able to see anything at all at the receiver. Packets at a higher layer enhance the quality of the received video. Thus, low-speed receivers would subscribe to a few lower layers, and high-speed receivers can enhance their reception quality by subscribing to higher layers, in addition to the lower layers. In general, if a link is shared by many users and the transmission rates and the number of users vary with time, the available rate on a link may vary. Thus, a control algorithm is necessary for a receiver to adjust its received rate (by varying the number of video layers to which it subscribes), depending upon the level of congestion in the network. In this section, we will extend the results of the previous sections to cover the case of multirate, multicast algorithms.

Let R_s be the set of receivers corresponding to any session $s \in S$. In the case of unicast sessions, R_s is a singleton set, whereas, for multicast sessions, $|R_s| > 1$. We will use the term *virtual session* to indicate the connection from a multicast source to one of its receivers. Thus, there are R_s virtual sessions corresponding to each session. We use the notation (s, r) to denote a virtual session corresponding to session s. Let L_{sr} be the set of all links in the route of a virtual session (s, r). Let S_l be the set of all sessions passing through link l and let V_{sl} be the set of all virtual sessions of the session s using link l.

Let $x_{sr}(t)$ denote the rate at which the virtual session (s, r) sends data at time t and let $U_{sr}(x_{sr})$ be the utility that it derives from this data rate. The total flow in link l at time t is given by

$$\sum_{s \in S_l} \max_{(s,r) \in V_{sl}} x_{sr}(t).$$

The objective is to find rates for which the total utility is maximized. Thus, as in the unicast case, the problem can be posed as the following optimization problem.

$$\max \sum_{s \in S} \sum_{r \in R_s} U_{sr}(x_{sr}) \tag{3.23}$$

subject to

$$\sum_{s \in S_l} \max_{(s,r) \in V_{sl}} x_{sr} \le C_l, \qquad \forall l \in L, \tag{3.24}$$

$$x_{sr} \ge 0, \qquad \forall s \in S, \ r \in R_s. \tag{3.25}$$

The max functions in the total rate of a multicast session passing through a link poses a challenge in studying the rate control equations and their stability. One possible approach could be to split the constraints into multiple constraints to make the constraints linear. To see this, suppose a link A has two receivers of a multicast session passing through it with rates x_{01} and x_{02}. Let link A also have a unicast session with rate x_1 passing through it. Note that the link constraint for link A,

$$\max(x_{01}, x_{02}) + x_1 \le C_A ,$$

can be equivalently written as

$$x_{01} + x_1 \le C_A \quad \text{and} \quad x_{02} + x_1 \le C_A .$$

Here C_A is the capacity of the link. We can thus split the link rate constraint into multiple linear constraints depending on the number of virtual sessions using the link. Suppose that we define a penalty function in a similar vein as in the solution to the unicast problem, and define the rate control mechanism of each of the virtual sessions using the partial derivative of the penalty function. It is easy to see that this would require each link to keep track of the rates of the individual virtual sessions (since each virtual session passing through a link will impose an additional constraint for that link) and thus, this approach is not scalable. To address this problem, we start by approximating the multicast rate control problem from which we can find a stable, decentralized solution for the approximate problem. We will then use this as a heuristic to solve the precise multicast congestion control problem.

Observe that the function $\max_i(x_i)$ can be approximated by the function $(\sum_i x_i^n)^{\frac{1}{n}}$ for large enough n, when all x_i's are non-negative real numbers. We now consider the following constraints, where the max functions in (3.24) are replaced by their differentiable approximations.

$$\sum_{s \in S_l} \left(\sum_{(s,r) \in V_{sl}} x_{sr}^n \right)^{\frac{1}{n}} \leq C_l, \qquad \forall l \in L,$$

$$x_{sr} \geq 0, \qquad \forall s \in S,\ r \in R_s. \qquad (3.26)$$

Next, consider a penalty function formulation of the optimization problem (3.23) subject to (3.26) in which we maximize the following objective.

$$V_n(\mathbf{x}) = \sum_{s \in S} \sum_{r \in R_s} U_{sr}(x_{sr}) - \sum_{l \in L} \int_0^{(\sum_{m \in S_l} (\sum_{(m,j) \in V_{ml}} x_{mj}^n)^{\frac{1}{n}})} p_l(u) du. \qquad (3.27)$$

Further, consider the following congestion control algorithm to maximize the cost function given by (3.27).

$$\dot{x}_{sr} = \kappa_{sr}(x_{sr}) \left(U'_{sr}(x_{sr}(t)) - \sum_{l \in L_{sr}} p_l(\tilde{y}_l) G^{(n)}_{l,(sr)}(\mathbf{x}) \right), \qquad (3.28)$$

where

$$\tilde{y}_l = \sum_{m \in S_l} (\sum_{(m,j) \in V_{ml}} x_{mj}^n(t))^{\frac{1}{n}},$$

and

$$G^{(n)}_{l,(sr)}(\mathbf{x}) = x_{sr}^{n-1} \sum_{(s,j) \in V_{s,l}} x_{sj}^n(t))^{\frac{1}{n}-1}.$$

We now show that the rate control differential equation given by (3.28) is stable for large but finite n, and the function $V_n(\mathbf{x})$ provides a Lyapunov function for the system of differential equations described in (3.28). Here $\mathbf{x}$ denotes the column vector of rates of all the virtual/unicast sessions.

Theorem 3.10. *The function $V_n(.)$ is a Lyapunov function for the system described in (3.28) and this system is globally asymptotically stable. The stable point is the unique $\mathbf{x}$ maximizing the strictly concave function $V_n(.)$.*

Proof. If we show $V_n(.)$ is a concave function, then the result follows along the lines of earlier proofs in this chapter. The functions $U_{sr}(.)$'s are strictly concave functions. Thus to show the convexity of $V_n(.)$ it suffices to show the convexity of the second term on the right-hand side of the expression of $V_n(.)$. Let

$$h_l(\mathbf{x}) = \sum_{m \in S_l} \left(\sum_{(m,j) \in V_{ml}} x_{mj}^n \right)^{\frac{1}{n}}$$

It is easy to see that the function $h_l(.)$ is convex. Also, since $p_l(.)$ is a monotonically non-decreasing function, the function $\int_0^x (p_l(u)du)$ is convex in x. Thus,

$$\int_0^{h_l(\mathbf{y})} (p_l(u)du) - \int_0^{h_l(\mathbf{x})} (p_l(u)du) \geq (h_l(\mathbf{y}) - h_l(\mathbf{x}))(p_l(h_l(\mathbf{x}))$$

$$\geq (p_l(h_l(\mathbf{x}))(\nabla h_l(\mathbf{x}))^T (\mathbf{y} - \mathbf{x}).$$

The second inequality follows from the convexity of $h_l(.)$. From this we can see that $\int_0^{h_l(\mathbf{x})}(p_l(u)du)$ is convex in $\mathbf{x}$ and so $V_n(.)$ is strictly concave. $\square$

The stability of (3.28) for arbitrarily large n thus motivates us to suggest a set of rate control equations for multicast networks. First, observe the following.

$$\lim_{n \to \infty} G_{l,(sr)}^{(n)}(\mathbf{x}) =$$
$$\begin{cases} 0, & \text{if } (s,r) \in V_{sl} \text{ and } x_{sr} < \max_{(s,j) \in V_{sl}}(x_{sj}), \\ 1, & \text{if } (s,r) \in V_{sl}, \ x_{sr} = \max_{(s,j) \in V_{sl}}(x_{sj}), \\ & \text{and } \forall \ (j \neq r, (s,j) \in V_{sl}) \ x_{sr} > x_{sj}, \\ (\sum_{(s,j) \in V_{sl}} I_{(x_{sr}=x_{sj})})^{-1} & \text{if } (s,r) \in V_{sl}, \ x_{sr} = \max_{(s,j) \in V_{sl}}(x_{sj}), \\ & \exists \ (j \neq r, (s,j) \in V_{sl}) \text{ st } x_{sr} = x_{sj}. \end{cases}$$

This suggests the following rate control equation for the virtual sessions in a multicast network.

$$\dot{x}_{sr} = \kappa_{sr}(x_{sr}) \left(U_{sr}'(x_{sr}) - \sum_{l \in L_{sr}} p_l(y_l) \frac{I_{(x_{sr}=\max_{(s,j) \in V_{sl}}(x_{sj}))}}{\sum_{(s,j) \in V_{sl}} I_{(x_{sr}=x_{sj})}} \right), \quad (3.29)$$

where we recall that y_l is the arrival rate at link l. The above rate control equation suggests that a link price is added to the path price of a virtual session only if that virtual session has a maximum rate among all the virtual sessions of its multicast session passing through that link. When more than one multicast receiver has a rate equal to the multicast session rate through the link, the link price should be split equally among all the multicast receivers with rate equal to the multicast session rate. So, from the point of view of a virtual session, it does not have to react to congestion on those links in which its rate is less than the multicast session rate.

To establish the stability of the congestion control mechanism given in (3.29), consider the function

$$V(\mathbf{x}) = \sum_{s \in S} \sum_{r \in R_s} U_{sr}(x_{sr}) - \sum_{l \in L} \int_0^{(\sum_{m \in S_l}(\max_{(m,j) \in V_{ml}} x_{mj})} p_l(u)du.$$

Lemma 3.11. *The function* $V(\mathbf{x})$ *is strictly concave in* $\mathbf{x}$.

Proof. This follows from the concavity of the utility functions, the convexity of the max(.) functions and the fact that the marking functions are increasing functions of the total rates. □

The congestion control equation (3.29) can be written as

$$\dot{x}_{sr} = \kappa_{sr}(x_{sr})(\nabla_s V(\mathbf{x}(t)))_{sr}, \qquad (3.30)$$

where $\nabla_s V(\mathbf{x}(t))$ is a particular sub-gradient[1] and $(\nabla_s V(\mathbf{x}(t)))_{sr}$ denotes the (s, r)-th component of the sub-gradient. If the point $\hat{\mathbf{x}}$ lies on the hyperplanes of non-differentiabilities of $V(.)$, then the particular sub-gradient we are using in the algorithm need not have a zero. For proving the convergence of rate control equation given in (3.29) we make the following assumptions.

Assumption 3.12 *The strictly concave function* $V(\mathbf{x})$ *attains its maximum at a point where its gradient exists.* □

The preceding assumption is required for the proof of stability. Suppose the maximum of the function $V(\mathbf{x})$ lies on a hyperplane of non-differentiability. Since the gradient of $V(\mathbf{x})$ does not exist at the minimum point, there exists a sub-gradient of $V(\mathbf{x})$ which is zero at the minimum point, but this need not correspond to the sub-gradient used in the right-hand side of the rate control as shown in (3.30). However, under Assumption 3.12, $V(\mathbf{x})$ reaches its maximum at a point where its gradient exists and so the equilibrium point of the rate control mechanism, $\hat{\mathbf{x}}$, also corresponds to the maximum of $V(\mathbf{x})$.

Theorem 3.13. *The system of multicast rate control equations given by (3.29) is globally asymptotically stable.*

Proof. First define the candidate Lyapunov function

$$W(\mathbf{x}) = \sum_{s \in S} \sum_{r \in R_s} \int_{x_{sr}^*}^{x_{sr}} \frac{1}{\kappa_{sr}(u_{sr})}(u_{sr} - x_{sr}^*)du_{sr}.$$

---

[1] The sub-gradient for a convex function can be viewed as a generalized gradient even when the gradient does not exist. For "well behaved" functions any convex combination of all the gradients around a small neighborhood of the point where the gradient does not exist can serve as a sub-gradient. Wherever the gradient exists, the sub-gradient is unique and is same as the gradient [9]. See Appendix 2.3.

It is easy to see that $W(\mathbf{x})$ is a radially unbounded (see Appendix 3.10 for the definition of radially unbounded) positive function with a unique zero at $\hat{\mathbf{x}}$. We also denote the gradient of $W(\cdot)$ by $\nabla W(\cdot)$. For $\mathbf{x} \neq \hat{\mathbf{x}}$, note the following for $\delta > 0$:

$$W(\mathbf{x}(t+\delta)) - W(\mathbf{x}(t))$$
$$= W(\mathbf{x}(t) + \delta\dot{\mathbf{x}}(t) + o(\delta)) - W(\mathbf{x}(t))$$
$$= (\delta\dot{\mathbf{x}}(t) + o(\delta))^t \nabla W(\mathbf{x}(t)) + o(\delta)$$
$$= \delta \sum_{s \in S} \sum_{r \in R_s} (x_{sr}(t) - \hat{x}_{sr})(\nabla_s V(\mathbf{x}(t)))_{sr} + o(\delta)$$
$$= \delta(\mathbf{x}(t) - \hat{\mathbf{x}})^t (\nabla_s V(\mathbf{x}(t))) + o(\delta).$$

Since $\nabla_s V$ is the sub-gradient of a strictly concave function, we have

$$V(\mathbf{x}(t)) - V(\hat{\mathbf{x}}) > (\mathbf{x}(t) - \hat{\mathbf{x}})^t (\nabla_s V(\mathbf{x}(t)))$$

from which it follows that,

$$W(\mathbf{x}(t+\delta)) - W(\mathbf{x}(t)) < \delta(V(\mathbf{x}(t)) - V(\hat{\mathbf{x}})) + o(\delta) .$$

We thus have,

$$D^+(W(\mathbf{x}(t))) = \limsup_{\delta \to 0^+} \frac{W(\mathbf{x}(t+\delta)) - W(\mathbf{x}(t))}{\delta}$$
$$< \limsup_{\delta \to 0^+} \frac{\delta(V(\mathbf{x}(t)) - V(\hat{\mathbf{x}})) + o(\delta)}{\delta}$$
$$= (V(\mathbf{x}) - V(\hat{\mathbf{x}})) < 0.$$

The last inequality follows from the fact that, under Assumption 3.12, $V(.)$ attains its maximum at $\hat{\mathbf{x}}$. In the above, D^+ denotes the upper right Dini derivative[2] with respect to the time variable. It is also easy to see that $D^+(W(\hat{\mathbf{x}})) = 0$. Hence the system is globally asymptotically stable and all trajectories ultimately converge to $\hat{\mathbf{x}}$.   □

3.9 A pricing interpretation of proportional fairness

Consider a network which allocates resources in the following manner. Each user r in the network bids a certain amount w_r, which indicates the dollars-per-sec that it is willing to pay. The network then allocates bandwidth to the users according to a weighted proportionally-fair rule with the weights being $\{w_r\}$ and charges q_r dollars-per-unit-bandwidth for the service. What should

---

[2] The upper-right Dini derivative of a function $f(.)$ at t is defined as $D^+(f(t)) = \limsup_{\delta \to 0^+} \frac{f(t+\delta) - f(t)}{\delta}$.

q_r be for this service to make sense? To answer this, recall that a weighted proportional fair service allocates $\{x_r\}$ to maximize the following objective:

$$\sum_r w_r \log x_r,$$

subject to

$$\sum_{s:s \in l} x_s \leq c_l, \quad \forall l,$$

and $x_r \geq 0$, $\forall r$. Suppose that q_r is chosen to be the sum of the Lagrange multipliers corresponding to the links on user r's path; then from the Karush-Kuhn-Tucker conditions, we have

$$w_r = q_r x_r.$$

Since w_r is dollars-per-sec, q_r is dollars-per-bit-per-sec and x_r is bits/sec, the amount that the user is willing to pay exactly matches the amount that the network charges. Thus, one can view this as a reasonable pricing scheme.

Now, suppose we ask a different question: how much would a rational user r bid in a such a network? In other words, what is the optimal value of w_r in such a network? To answer this question, we have to model the user's utility when it receives a certain amount of bandwidth from the network. Suppose that $U_r(x_r)$ is the utility function of user r and suppose that it is measured in dollars-per-sec. Thus, when quoted a price q_r by the network, the user knows that it will get w_r/q_r bits-per-sec if it bids w_r dollars-per-sec and therefore, its utility will be $U_r(w_r/q_r)$. Then, a natural optimization problem for user r is given by

$$\max_{w_r \geq 0} U_r\left(\frac{w_r}{q_r}\right) - w_r.$$

Suppose we make the reasonable assumption that user r cannot anticipate the impact of its bid w_r on the price q_r, i.e., q_r is not a function of w_r from user r's point of view. The reason that this assumption is reasonable is that, in a large network, it is impossible for each user to assess the impact of its own action on the network's pricing strategy. Assuming that $U_r(x_r) \to \infty$ as $x_r \to 0$, the optimal value of w_r is strictly greater than zero and is given by

$$U_r'\left(\frac{w_r}{q_r}\right) = q_r.$$

Note that the user response can be equivalently written in two other forms:

$$q_r = U_r'(x_r) \tag{3.31}$$

and

$$w_r = x_r U_r'(x_r). \tag{3.32}$$

Let us look at the network from another point of view: one of allocating resources in a fair manner among competing users. Recall that our notion of fairness is to optimize the sum of the utility functions of all the users in the network, i.e.,

$$\max_{\{x_r\}} \sum_r U_r(x_r).$$

It is interesting to ask the following question: is there a set of $\{w_r\}$, $\{q_r\}$ and $\{x_r\}$ such that the following objectives are met?

- Given $\{w_r\}$, the optimal user bandwidths under weighted proportional fairness are $\{x_r\}$ and the corresponding prices are $\{q_r\}$.
- Given $\{q_r\}$, the optimal user bids are $\{x_r\}$.
- Given $\{U_r(\cdot)\}$, the fair bandwidth allocation is $\{x_r\}$.

The answer is yes since the proportionally-fair pricing/bidding mechanism also ensures that (3.31) is satisfied, which is the Karush-Kuhn-Tucker condition for the fair resource allocation problem. This pricing interpretation was first given in [48] for the case for each r, $U_r(x_r) \rightarrow \infty$ when $x_r \rightarrow 0$. The more general case, not considered in this book, is analyzed in [42].

The bidding/pricing interpretation can also be carried out for the congestion controller that leads to the solution of the weighted proportionally-fair resource allocation. Recall that one version of the proportionally-fair primal congestion controller is given by

$$\dot{x}_r = \kappa_r \left(w_r - x_r q_r \right),$$

where $\kappa_r > 0$ is a constant,

$$q_r = \sum_{l:l \in r} f_l(y_l)$$

and

$$y_l = \sum_{s:l \in s} x_s.$$

Further, recall that $f_l(y_l)$ is the penalty for using link l when the arrival rate at the link is y_l. Suppose that the network charges $f_l(y_l)$ dollars-per-bit-per-sec for using link l. Then, at equilibrium, since

$$w_r = x_r q_r,$$

the user is charged w_r dollars-per-sec for the service he/she receives from the network. If the user chooses w_r according to (3.32), then the dynamics of the congestion controller is given by

$$\dot{x}_r = \kappa_r \left(x_r U_r'(x_r) - x_r q_r \right),$$
$$= \kappa_r x_r \left(U_r'(x_r) - q_r \right),$$

which was proved to lead to fair resource allocation in Theorem 3.4. An alternate which allows the users to compute their own prices without explicit feedback from the network is presented in [64].

3.10 Appendix: Lyapunov stability

In this appendix, we present some results on Lyapunov stability theory that were used in this chapter and will be useful in subsequent chapters. We only provide a quick overview here and the interested reader is referred to [55] for further details.

Consider a differential equation

$$\dot{\mathbf{x}} = f(\mathbf{x}), \qquad \mathbf{x}(0) = \mathbf{x}_0, \qquad (3.33)$$

where $f(\mathbf{x})$ is assumed to satisfy appropriate conditions such that the differential equation has a unique solution. Assume that $\mathbf{0}$ is the unique equilibrium point of (3.33), i.e.,

$$f(\mathbf{x}) = 0$$

has a unique solution given by $\mathbf{x} = \mathbf{0}$. We now define various notions of the stability of (3.33).

Definition 3.14. *We define the following notions of stability of the equilibrium point:*

- *The equilibrium point is said to be stable if, given an $\varepsilon > 0$, there exists a $\delta > 0$, such that*

$$\|\mathbf{x}(t)\| \le \varepsilon, \quad \forall t \ge 0, \qquad if \; \|\mathbf{x}_0\| \le \delta.$$

- *The equilibrium point is asymptotically stable if there exists a $\delta > 0$ such that*

$$\lim_{t \to \infty} \|\mathbf{x}(t)\| = 0$$

for all $\|x_0\| \le \delta$.
- *The equilibrium point is globally, asymptotically stable if*

$$\lim_{t \to \infty} \|\mathbf{x}(t)\| = 0$$

for all initial conditions $\|x_0\|$. □

Intuitively, stability implies that if the trajectory of the solution of the differential equation is perturbed by a small amount from the equilibrium, then the trajectory still stays around the equilibrium point. The definition of asymptotic stability states that the trajectory eventually reaches the equilibrium point. Global asymptotic stability strengthens this further by requiring asymptotic stability starting from any initial condition. The Lyapunov theorem, which is stated next, gives a sufficient condition to check for stability.

Theorem 3.15. *Consider a continuously differentiable function $V(\mathbf{x})$ such that*

$$V(\mathbf{x}) > 0, \qquad \forall \mathbf{x} \neq \mathbf{0}$$

and $V(\mathbf{0}) = 0$. Now we have the following conditions for the various notions of stability.

1. *If $\dot{V}(\mathbf{x}) \leq 0 \;\forall \mathbf{x}$, then the equilibrium point is stable.*
2. *In addition, if $\dot{V}(\mathbf{x}) < 0, \;\forall \mathbf{x} \neq \mathbf{0}$, then the equilibrium point is asymptotically stable.*
3. *In addition to (1) and (2) above, if V is radially unbounded, i.e.,*

$$V(\mathbf{x}) \to \infty, \qquad when \; \|\mathbf{x}\| \to \infty,$$

   then the equilibrium point is globally asymptotically stable.

$\square$

It is useful to get an intuitive understanding of Lyapunov's theorem. Consider a region Ω_c defined as

$$\Omega_c = \{\mathbf{x} : \|V(x)\| \leq c\}.$$

Since $\dot{V} \leq 0$, every trajectory starting in Ω_c will remain in Ω_c. Further, by making c small, we can make Ω_c small, thus ensuring stability. If $\dot{V}(\mathbf{x}) < 0$, for $\mathbf{x} \neq \mathbf{0}$, then the trajectory can be contained in a sequence of regions Ω_c where c becomes small as $t \to \infty$. However, this argument breaks down if Ω_c is an unbounded set. From the definition of stability, we need to find a region

$$B_\delta = \{\mathbf{x} : \|\mathbf{x}\| \leq \delta\}$$

such that the trajectory lies in this region. If Ω_c is unbounded for some $c < \infty$, then we cannot find a B_δ such that $\Omega_c \subset B_\delta$. Condition (3) in Theorem 3.15 ensures that, for every finite c, Ω_c is a bounded set.

We note that Lyapunov's theorem only provides sufficient conditions for stability. In other words, if we cannot find a $V(\mathbf{x})$ that satisfies the conditions of the theorem, we cannot conclude that the system is unstable.

Lyapunov's theorem can be generalized to situations where $V(\mathbf{x})$ may not be differentiable. In such cases, usually it is enough to verify that

$$D^+ V = \lim_{\delta \to 0^+} \frac{V(\mathbf{x}(t + \delta)) - V(\mathbf{x}(t))}{\delta}$$

satisfies the conditions on $\dot{V}$ in Theorem 3.15. In other words, it is sufficient to check the conditions of the Lyapunov theorem only in forward time. The reader is referred to [89] for further details. We used such a result in proving the stability of multicast congestion control algorithms.

4

Relationship to Current Internet Protocols

4.1 Window flow control

Consider a single source accessing a link which has the capacity to serve c packets-per-second. Let us also suppose for simplicity that all packets are of equal size. To ensure that congestion does not occur at the link, the source should transmit at a maximum rate of c. One way to ensure this is using a *window flow control* protocol. A source's *window* is the maximum number of unacknowledged packets that the source can inject into the network at any time.

For example, if the window size is 1, then the source maintains a counter which has a maximum value of 1. The counter indicates the number of packets that it can send into the network. The counter's value is initially equal to the window size. When the source sends one packet into the network, the counter is reduced by 1. Thus, the counter in this example would become zero after each packet transmission and the source cannot send any more packets into the network till the counter hits 1 again. To increment the counter, the source waits for the destination to acknowledge that it has received the packet. This is accomplished by sending a small packet called the *ack* packet, from the destination back to the source. Upon receiving the ack, the counter is incremented by 1 and thus, the source can again send one more packet. We use the term *round-trip time (RTT)* to refer to the amount of time that elapses between the instant that the source transmits a packet and the instant at which it receives the acknowledgment for the packet. With a window size of 1, since one packet is transmitted during every RTT, the source's data transmission rate is $1/RTT$ packets/sec.

If the window is 2, the counter's value is initially set to 2. Thus, the source can send two back-to-back packets into the network. For each transmitted packet, the counter is decremented by 1. Thus, after the first two packet transmissions, the counter is decremented to zero. When one of the packets is acknowledged and the ack reaches the source, then the source increments the counter by 1 and can send one more packet into the network. Once the new packet is transmitted, the counter is again decremented back to zero. Thus, after each ack, one packet is sent, and then, the source has to wait for the next ack before it can send another packet. If one assumes that the processing speed of the link is very fast, i.e., $1/c << RTT$, and that the processing times at the source and destination are negligible, then the source can transmit two packets during every RTT. Thus, the source's transmission rate is $2/RTT$ packets/sec. From the above argument, it should be clear that, if the window size is W, then the transmission rate can be approximated by W/RTT packets/sec. A precise computation of the rate as a function of the window is difficult since one has to take processing delays into account at the source and destination and the queueing delays at the link. It is not clear that such an effort would provide any insight into the operation of a high-speed network and therefore, in common with current literature, we will use the approximate relationship between the window and the transmission rate.

If the link capacity is c and the source's window size W is such that $W/RTT < c$, then the system will be stable. In other words, all transmitted packets will be eventually processed by the link and reach the intended destination. However, in a general network, the available capacity cannot be easily determined by a source. The network is shared by many sources which share the capacities at the various links in the network. Thus, each source has to adaptively estimate the value of the window size that can be supported by the network. The solution proposed for this by Jacobson [39] is the subject of the next section.

4.2 Jacobson's adaptive window flow control algorithm

Before we delve into the adaptive window flow control algorithm, it would be useful to understand the layered architecture of the Internet to understand the framework within which the adaptive window flow control protocol is implemented in the Internet. We will provide a very brief introduction here and refer the reader to [10] for a more detailed discussion. Roughly speaking, the Internet is organized in several layers:

- Physical layer
- Data link layer
- Network layer
- Transport layer
- Application layer.

The physical layer refers to the collection of protocols that are required to transmit a bit (a 0 or a 1) over a physical medium such as an ethernet cable. Typically, the physical medium accepts a waveform as an input and produces a waveform as the output. Thus, we need protocols to convert 0s and 1s into these waveforms. This function is performed at the physical layer.

The data link layer refers to the collection of protocols that collects many bits together in the form of a *frame* and ensures that the frame is delivered from one end of the physical link to the other. Error correction could also be added to ensure that errors in the frame transmission can be detected and corrected.

The network layer is the collection of protocols that is used to append end-host addresses and other information to data bits to form a packet and further, to use these addresses to route packets through the network. Thus, this layer performs the crucial task of routing or delivering a packet from a source to a destination. In the Internet, this layer is also called the IP (Internet Protocol) layer. While the network layer performs the packet delivery service, the packet delivery may not be reliable. In other words, packets could be lost or corrupted on the route from the source to the destination. One reason for loss of packets is the overflow of buffers at the routers in the network if the source transmission rates are larger than the rate at which packets can be processed by the routers.

The transport layer adds reliability to the network layer. The transport layer protocols ensure that lost packets are detected and possibly retransmitted from the source, if necessary, depending upon the application. The predominant transport layer protocol used in today's Internet is the Transmission Control Protocol (TCP). Jacobson's adaptive window flow control algorithm is implemented within TCP. TCP is primarily used by file transfer applications which need reliable, in-sequence delivery of packets from the source to the destination. There are other transport layer protocols such as UDP (User Datagram Protocol) which are used by applications such as video transmission which can tolerate some amount of packet loss, where packet retransmission may not be required.

Finally, the application layer consists of protocols such as ftp, http, etc. which use the lower layers to transfer files or other forms of data over the Internet.

The above discussion is not intended to be an in-depth or precise discussion of the layered architecture of the Internet. It is rather intended to serve as a quick overview aimed at pointing out the layer within which TCP is implemented in the Internet and further, to indicate that congestion control is implemented within the transport layer protocol TCP. It also serves to highlight the fact what we are studying in this book is just a very small part of the giant collection of protocols that make the Internet function.

We are now ready to describe Jacobson's algorithm implemented in today's TCP. TCP uses a scheme that changes its window size depending on an estimate of congestion in the network. The basic idea is that the window size is increased till buffer overflow occurs. The overflow is detected by the destination due to the fact that some of the packets do not reach the destination. Upon detecting lost packets, the destination informs the source which resets the window to a small value. When there is no packet loss, the window is increased rapidly when it is small, and then after the window size reaches a threshold, it is increased more slowly later by probing the network for bandwidth.

Jacobson's congestion control algorithm operates in two phases:

Slow-Start Phase:

- Start with a window size of 1.
- Increase the window size by 1 for every ack received. This continues till the window size reaches a threshold called the slow-start threshold (*ssthresh*). The initial value of *ssthresh* is set at the beginning of the TCP connection when the receiver communicates the maximum value of window size that it can handle. The initial value of *ssthresh* is set to be some fraction (say, half) of the maximum window size. Once the window size reaches *ssthresh*, the slow-start phase ends, and the next phase called the *congestion avoidance* begins. If a packet loss is detected before the window size reaches *ssthresh*, then *ssthresh* is set to half the current window size, then the current window size is reset to 1, and slow-start begins all over again.

Congestion Avoidance Phase:

- In the congestion avoidance phase, the window size is increased by

$$1/\lfloor cwnd \rfloor$$

for every ack received, where *cwnd* denotes the current window size. This is roughly equivalent to increasing the window size by 1 after every *cwnd* acks are received.
- When packet loss is detected, the window size is decreased. Different versions of TCP, such as TCP-Tahoe, TCP-Reno, TCP-NewReno, TCP-SACK, ,, reduce the window size in different ways. However, for our modelling purposes, these do not make much of a difference and therefore, we will assume the following algorithm: set *ssthresh* to half the current window size, *cwnd* to one and go back to the slow-start phase.

When a packet loss is detected, the lost packet and all the succeeding packets are retransmitted, i.e., selective retransmission is not implemented in most versions of TCP. The exception is TCP-SACK, where sometimes only the lost packets are retransmitted.

In the description of Jacobson's algorithm we did not specify how TCP detects loss. In the early versions of this algorithm, called TCP-Tahoe, loss was detected only if there was a timeout, i.e., an ack is not received within a certain amount of time. In more recent versions such as TCP-Reno and TCP-NewReno, packet loss is assumed either if there is a timeout or if three duplicate acks are received. Why do multiple acks for the same packet denote a loss? To understand this, we have to first understand how the destination generates an ack.

All TCP packets are given a sequence number. For the purposes of the discussion here, let us suppose that the first packet is given a sequence number 1, the second packet is numbered 2, and so on. When a packet reaches its destination, the destination generates an ack packet and puts the sequence number of the next packet that it expects to receive in the ack and sends it to the source. For example, suppose packets 1 through 9 have been received by the destination, and then packet 10 arrives at the destination. Instead of stating that it received packet 10, the ack corresponding to packet 10 will state that the next expected packet is 11. Now, if packet 11 is lost due to a buffer overflow inside the network or due to some other reason, then the destination receives packet 12. Now the ack for packet 12 would again say that the next expected packet is 11. This is because 11 is still missing and the destination expects to receive it before the TCP session has been completed. Now, suppose that packet 13 arrives at the destination, the destination generates yet another ack stating that it expects 11 next. At some point, the source has to realize that 11 was lost and prepare to both retransmit the packet as well as reduce the window size in response to possible congestion that led to the packet loss. In today's versions of TCP, it is assumed that when three acks are received at the source with the same expected next-packet sequence number, it is assumed that a packet is lost and this triggers the congestion control algorithm to reduce its window size.

We need an exception to this rule for detecting packet loss if all the remaining packets in a window are lost during a congestion episode. If all the packets are lost, then there are no further packets that will reach the destination to trigger duplicate acks. Thus, in this case, the source simply waits for a certain amount of time called the *timeout* interval and after this interval of time, if no acks are received, then a packet loss is assumed to have occurred.

To further understand the behavior of Jacobson's algorithm, we present the following simple analysis of a single TCP source accessing a single link that was presented in [65]. Suppose that c is the capacity of the link in packets per second and let τ denote the round-trip propagation delay. Thus, the RTT which is the sum of the propagation and queueing delays is given by $T = \tau + 1/c$. Suppose that the link capacity is 50 Mbps and packet size is 500 bytes, then $c = 12,500$ packets/sec. and $1/c = 0.08$ msecs. If the transmission is taking place over a distance of 1000 km, using the fact that the speed of light is 3×10^8 m/sec. and ignoring the refractive index of the transmission medium, round-trip propagation delay $\tau = 2 \times 10^6 / 3 \times 10^8 = 6.6$ msecs.

The product cT is called the *delay-bandwidth* product. In high-speed networks, this quantity is large since c is large. Let B denote the buffer size at the link. We will assume that $B \leq cT$. Suppose that the buffer is non-empty, then packets are processed at rate c by the link. Thus, acks are generated by the destination at rate c and therefore, new packets can be released by the source every $1/c$ seconds so that there is at least one space in the buffer for each new packet. For simplicity, we assume that the propagation delay from the source to the link is zero and the propagation delay from the destination back to the source is τ. Since the propagation delay is τ, $c\tau$ acks are in transit. Since there are B packets are in the buffer, the total number of unacknowledged packets is $cT + B$. If the window size is larger than this, then buffer overflow would occur. Thus, the maximum possible window size without leading to buffer overflow is $W_{max} = cT + B$.

When a packet loss does occur, the window size will be slightly larger than W_{max} and the amount by which it is larger depends on c and the RTT. After a packet loss, *ssthresh* is set to half the current window size which is larger than W_{max}. As an approximation, we will assume that *ssthresh* is equal to $(\mu T + B)/2$.

Time	Packet Ack'ed	Window Size	Packet Released	Queue Length
Mini-Cycle 0				
0		1	1	1
Mini-Cycle 1				
T	1	2	2,3	2
Mini-Cycle 2				
2T	2	3	4,5	2
2T+1/c	3	4	6,7	2-1+2=3
Mini-Cycle 3				
3T	4	5	8,9	2
3T+1/c	5	6	10,11	2-1+2=3
3T+2/c	6	7	12,13	2-1+2-1+2=4
3T+3/c	7	8	14,15	2-1+2-1+2=5
Mini-Cycle 4				
4T	8	9	16,17	2
$\vdots$				

Table 4.1. Slow-start phase

The evolution of the window size and queue length during the slow-start phase is shown in Table 4.1. A mini-cycle refers to the time it takes for the window size to double. At time $t = 0$, the first packet is released by the source. It takes one RTT, equal to T time units, for this packet's ack to reach the source. When the ack is released the window size is increased by 1, i.e., it is incremented from 1 to 2. This starts the next mini-cycle. At this time, the source releases two packets into the network. It takes one RTT for the first of these packets to be acknowledged. The window size becomes 3 at this time, and since there is only one unacknowledged packet in the network, the source releases two more packets into the network. Further evolution of the window size is shown in Table 4.1.

From the table, it is easy to see that the window size satisfies the following equation: for $n \geq 1$,

$$W(nT + m/c) = 2^{n-1} + m + 1, \ 0 \le m \le 2^{n-1} - 1.$$

The queue length satisfies

$$Q(nT + m/c) = m + 2, \ 0 \le m \le 2^{n-1} - 1.$$

Note that the maximum window size and the maximum queue length in a mini-cycle satisfy

$$Q_m := \text{Max queue length in a mini cycle} = 2^{n-1} + 1$$

and

$$W_m := \text{Max window size in a mini cycle} = 2^n.$$

Note that $Q_m \approx W_m/2$.

Buffer overflow does not occur in the slow-start phase if $Q_m \le B$. But $Q_m \approx W_m/2 \le (cT + B)/4$. Thus, a sufficient condition for no buffer overflow in slow start is

$$(cT + B)/4 \le B,$$

which is equivalent to

$$B \ge cT/3.$$

If $B < cT/3$, then buffer overflow does occur since the queue length when the window size is $W_{max}/2$ is $W_{max}/4 = (cT + B)/4$ which is greater than B. Therefore, we consider two separate cases: one where the buffer does not overflow in the slow-start phase and the other where it does.

Case I: $B > cT/3$.

We first note that $W(t) \approx 2^{t/T}$. Thus, the length of the slow-start phase, T_{ss}, is given by

$$2^{T_{ss}/T} = (cT + B)/2,$$

or

$$T_{ss} = T \log_2 \left(\frac{cT + B}{2} \right).$$

The number of packets successfully transmitted (i.e., those for which acks were received) during slow-start is given by

$$N_{ss} = \frac{cT + B}{2},$$

since the window size increases by 1 for every received ack.

Case II: $B < \mu T/3$.

There are two slow-start phases since buffer overflow occurs in the first slow-start phase. In the first slow-start phase, buffer overflow occurs when the queue length exceeds B. Since the window size is approximately twice the queue length, buffer overflow occurs when the window size is approximately $2B$. Thus, the duration of the first slow-start phase is

$$T_{ss1} = T \log_2 2B + T,$$

where the additional T is due to the fact that it takes one round-trip time to detect the loss. As before,

$$N_{ss1} = 2B.$$

Since the window size roughly doubles every T units of time, the window size when packet loss is detected can be shown to be $\min\{4B - 2, ssthresh\}$, where $ssthresh = (cT + B)/2$. This implies that the second slow-start phase uses $ssthresh = \min\{2B - 1, (cT + B)/4\}$. Thus,

$$T_{ss2} = T \log_2 \min\{2B - 1, (cT + B)/4\},$$

and

$$N_{ss2} = \min\{2B - 1, (cT + B)/4\}.$$

Overall,

$$T_{ss} = T_{ss1} + T_{ss2},$$

and

$$N_{ss} = N_{ss1} + N_{ss2}.$$

Next, we turn our attention to the congestion avoidance phase. The starting window size for the congestion avoidance phase is given by

$$W_0 = \begin{cases} W_{max}/2 & \text{if } B > cT/3, \\ \min\{2B - 1, (cT + B)/4\} & \text{if } B < cT/3. \end{cases}$$

We will derive a differential equation to describe the window evolution in the congestion avoidance phase. Let $a(t)$ denote the number of acks received by the source after t units of time into the congestion avoidance phase. We can express $\frac{dW}{dt}$ as

$$\frac{dW}{dt} = \frac{dW}{da}\frac{da}{dt}.$$

If the window size is large enough to keep the server busy continuously, then the rate at which acks are received is c. Otherwise, the rate at which acks are received is W/T. Thus,

$$\frac{da}{dt} = \min\{W/T, c\}.$$

Recall that, in the congestion avoidance phase, the window size is increased by $1/W$ for every received ack. Therefore,

$$\frac{dW}{da} = 1/W,$$

and

$$\frac{dW}{dt} = \begin{cases} 1/T & \text{if } W \leq cT, \\ c/W & \text{if } W > cT. \end{cases}$$

Thus, the congestion avoidance phases consists of two sub-phases corresponding to $W \leq cT$ and $W > cT$. The duration of the first phase is

$$T_{ca1} = T(cT - W_0),$$

and the number of packets successfully transmitted in this phase is

$$\begin{aligned} N_{ca1} &= \int_0^{T_{ca1}} a(t)dt \\ &= \int_0^{T_{ca1}} \frac{W(t)}{T} dt \\ &= \int_0^{T_{ca1}} \frac{W_0 + t/T}{T} dt \\ &= \frac{W_0 T_{ca1} + T_{ca1}^2/(2T)}{T}. \end{aligned}$$

When $W > cT$, $W(t)^2/2 = ct + (\mu T)^2/2$, and thus,

$$T_{ca2} = \frac{W_{max}^2 - (\mu T)^2}{2\mu},$$

and

$$N_{ca2} = \mu T_{ca2},$$

since the node is fully utilized during this sub-phase. The total duration of the congestion avoidance phase T_{ca} and the total number of packets transferred from the source to the destination during the congestion avoidance phase N_{ca} are given by

$$T_{ca} = T_{ca1} + T_{ca2}, \quad N_{ca} = N_{ca1} + N_{ca2}.$$

From the above analysis, we can compute the durations of the slow-start and congestion-avoidance phases as well as the total number of packets transmitted in each phase. Thus,

$$\text{TCP Throughput} = \frac{N_{ss} + N_{ca}}{T_{ss} + T_{ca}}.$$

Consider a bottleneck link with data rate 1 Gbps. Assume that each packet is of size 1000 bytes, and let the source and destination be 1000 km apart. We will now compute the throughput of Jacobson's algorithm assuming a buffer size of 400 packets. Since the speed of light is 3×10^8 m/s, the round-trip propagation delay is given

$$\tau = (2 \times 10^6)/(3 \times 10^8) = 6.67 \text{ msecs.}$$

The link capacity c is given by

$$c = 10^9/1000/8 = 125,000 \text{ packets/sec.}$$

Since $B = 400$, we have $B \geq cT/3$. Therefore, this corresponds to the case where the buffer is large enough for there to be only one slow-start phase before congestion avoidance begins. From the previous discussion

$$T_{ss} = T \log_2 \left(\frac{\mu T + B}{2} \right) = 0.0618 \text{ secs.,}$$

$$N_{ss} = \frac{\mu T + B}{2} = 616.9,$$

$$T_{ca1} = T(\mu T - \frac{\mu T + B}{2}) = 1.44639 \text{ secs.,}$$

$$N_{ca1} = 157,738,$$

$$T_{ca2} = 3.308,$$

and

$$N_{ca2} = 413,500.$$

Therefore,

$$\text{TCP Throughput} = \frac{413,500 + 157,738 + 617.4}{3.308 + 1.44639 + 0.0618} = 118,736 \text{ packets/sec.}$$

Notice that if we only consider the congestion avoidance phase,

$$\text{TCP Throughput} \approx \frac{413,500 + 157,738}{3.308 + 1.44639} = 120,150 \text{ packets/sec.}$$

Thus, we see that most of the bits are transferred during the congestion avoidance phase.

Next, we consider the same example with the buffer size $B = 200$. Then $B = 200$, $B < cT/3$, and using the previously derived expressions, we have

$$T_{ss1} = 0.064,$$

$$N_{ss1} = 400,$$

$$W_0 = 70.84,$$

$$T_{ss2} = 0.041,$$

$$N_{ss2} = 70.84,$$

$$T_{ca1} = 5.089,$$

$$N_{ca1} = 345,061,$$

$$T_{ca2} = 1.494,$$

$$N_{ca2} = 186,750.$$

Therefore,

$$\text{TCP Throughput} = \frac{400 + 70.84 + 345,061 + 186,750}{0.064 + 0.041 + 5.089 + 1.494} = 79,587 \text{ packets/sec.}$$

Notice that the throughput is severely degraded due to the fact that the buffer size is small, thus forcing two slow-start phases. The throughput obtained by considering the congestion avoidance phase alone is given by

$$\text{TCP Throughput} \approx \frac{345,061 + 186,750}{5.089 + 1.494} = 80,786.$$

Again, we see that most of the bits are transferred during the congestion avoidance phase. These two examples suggest that the slow-start phase lasts for an extremely short time compared to congestion avoidance; however, the buffer size may impact the duration of the slow-start phase and thus, have a significant impact on throughput. Note that, since the slow-start phase is very short compared to the congestion avoidance phase, the congestion-control behavior of TCP can be viewed as follows: increase the window size by 1 during each RTT and decrease the window size by half upon detecting a packet loss. Such an algorithm is called an *additive increase, multiplicative decrease* (AIMD) algorithm.

4.2.1 TCP and resource allocation

The model considered in the previous subsection is somewhat simplistic in the sense that it ignores the impact of random phenomena in the network. However, the model was useful in understanding the roles of slow-start and congestion avoidance. It also showed that the congestion avoidance phase is the dominant phase in the sense that the bulk of the data transfer by the source occurs in this phase. Therefore, in this subsection, we consider only the congestion phase and present a model for TCP behavior which relates the TCP dynamics to the resource allocation problem and the resulting gradient-procedure-like congestion control algorithms that were used to solve the resource allocation problem.

Let $W_r(t)$ denote the window size of flow r, T_r its RTT, and let $q_r(t)$ be the fraction of packets lost at time t. Then, the congestion avoidance phase of TCP-Reno can be modelled as

$$\dot{W}_r = \frac{x_r(t - T_r)(1 - q_r(t))}{W_r} - \beta x_r(t - T_r)q_r(t)W_r(t), \tag{4.1}$$

where $x_r(t) = W_r(t)/T_r$ is the transmission rate. The parameter β is the decrease factor and is taken to be $1/2$ although studies show that a more precise value of β when making a continuous-time approximation of TCP's behavior is 2/3 [62]. Substituting for $W_r(t)$ in terms of $x_r(t)$ gives

$$\dot{x}_r = \frac{x_r(t - T_r)(1 - q_r(t))}{T_r^2 x_r} - \beta x_r(t - T_r)q_r(t)x_r(t). \tag{4.2}$$

The equilibrium value of x_r is easily seen to be

$$\hat{x}_r = \sqrt{\frac{1 - \hat{q}}{\beta \hat{q}}} \frac{1}{T_r},$$

where $\hat{q}_r$ is the equilibrium loss probability. For small values of $\hat{q}_r$,

$$\hat{x}_r \propto 1/T_r\sqrt{\hat{q}_r},$$

which is a well-known result on the steady-state behavior of TCP-Reno [39]. If $T_r = 0$, (4.2) reduces to

$$\dot{x}_r = \frac{1 - q_r}{T_r^2} - \beta x_r^2 q_r$$
$$= (\beta x_r^2 + \frac{1}{T_r^2})\left(\frac{1}{\beta T_r^2 x_r^2 + 1} - q_r\right).$$

Comparing it to (3.8), we see that the derivative of the source's utility function is given by

$$U_r'(x_r) = \frac{1}{\beta T_r^2 x_r^2 + 1}.$$

Thus, the source utility function can be determined up to an additive constant as

$$U_r(x_r) = \frac{\arctan(x_r T_r \sqrt{\beta})}{\sqrt{\beta} T_r}.$$

Note that this function does not satisfy the condition that $U_r(x_r) \to -\infty$ as $x_r \to \infty$. Thus, to use Theorem 3.4, we have to assume that the solution to (3.5) is bounded away from zero.

If $q_r(t)$ is small, then $1 - \hat{q}_r \approx 1$, and the dynamics of x_r in the absence of feedback delay can be approximated as

$$\dot{x}_r = \frac{1}{T_r} - \beta x_r^2 q_r(t).$$

This corresponds to

$$U_r'(x_r) = \frac{1}{T_r x_r^2},$$

or the utility function can be determined to be

$$U_r(x_r) = -\frac{1}{x_r T_r},$$

up to an additive constant. Thus, when q is small, TCP solves a weighted minimum potential delay fair resource allocation problem.

4.3 TCP-Vegas

The schemes that we described in the previous section are variations of the following commonly implemented versions of TCP: Tahoe, Reno, NewReno, and SACK. They all rely on packet loss to measure congestion. Another measure of congestion is the queueing delay at the links. TCP-Vegas is a version of TCP that reacts to queueing delay [14]. The basic idea behind TCP-Vegas is as follows: if the delays are large, then decrease window size; else, increase window size.

The congestion control algorithm in TCP-Vegas has two parameters α and β, with $\alpha < \beta$. The TCP-Vegas congestion control algorithm operates as follows:

- *BaseRTT (Propagation Delay):* Assume that the source can somehow estimate the round-trip propagation delay, i.e., the total RTT minus the queueing delays. In TCP-Vegas, the propagation delay is denoted by the variable *BaseRTT*. The expected throughput is computed as

$$e = \frac{Transmission\ window\ size}{BaseRTT}.$$

The BaseRTT at time t is estimated to be the smallest RTT seen by the source up to time t. The idea is that occasionally a packet will encounter nearly empty queues. Thus, the smallest RTT seen so far is a good estimate of the propagation delay.

- *Actual Throughput:* In addition, the source computes the actual throughput as follows. It sends a distinguished packet and records the time at which it was transmitted. Then, it records the time at which the ack for the distinguished packet is received. The RTT is computed to be the difference between the ack reception time and the transmission time of the distinguished packet. During the RTT measured above, the source records the total number of received acks during this period. Then, the actual throughput is computed as

$$a = \frac{Number\ of\ acks\ received}{RTT}.$$

- TCP-Vegas compares e and a and adjusts the window size as a function of this difference between e and a. Note that $e \geq a$ unless the *BaseRTT* estimate is incorrect. Thus, if e turns out to be smaller than a, then *BaseRTT* is set to be equal to the RTT used to estimate e. If $e \geq a$, note that $(e - a)RTT$ packets transmitted in the previous RTT are still unacknowledged and therefore, are in the network. Depending upon the values of $(e - a)$ and e, one of the following congestion control actions is initiated:
  - If $\alpha \leq (e - a) \leq \beta$, then leave the window size unchanged. In this case, the congestion avoidance algorithm of TCP-Vegas concludes that the

actual throughput and the expected throughput are nearly equal and
thus, does not change the window size.
- If $(e - a) > \beta$, decrease the window size by 1 for the next RTT.
- If $(e - a) < \alpha$, increase the window size by 1 for the next RTT.

Consider the special case $\alpha = \beta$. In this case, the resource allocation
performed by TCP-Vegas can be interpreted in terms of the resource allocation
formulation in Chapter 2 if we make the following simplifying assumptions
about TCP-Vegas:

- TCP-Vegas estimates the propagation delay ($BaseRTT$) accurately. This
  can be done, for example, using the IP precedence protocol as in [98].
- The total RTT seen by source r is the sum of the propagation delay and
  the queueing delay on the route from the source r to its destination. In
  other words, any delays incurred by the ack packet is ignored.

Under these assumptions, the following theorem was proved in [70].

Theorem 4.1. *Let* $\alpha = \beta$ *in the TCP-Vegas congestion control algorithm and
suppose that a network of TCP-Vegas sources converge to an equilibrium and
the equilibrium rates are* $\{\hat{x}_r\}$. *Then,* $\{\hat{x}_r\}$ *is the solution to*

$$\max_{\{x_r\}} \alpha T_{pr} \log x_r, \qquad (4.3)$$

subject to

$$\sum_{r:l \in r} x_r \leq c_l,$$

and $x_r \geq 0$, $\forall r$, *where* T_{pr} *is the round-trip propagation delay of source* r.

Proof. Let W_r denote the equilibrium window size of source r and let T_{qr}
represent the equilibrium queueing delay on the route of source r. If the sources
reach an equilibrium, then the window size satisfies

$$\frac{W_r}{T_{pr}} - \frac{W_r}{T_{pr} + T_{qr}} = \alpha. \qquad (4.4)$$

If we make the approximation

$$\hat{x}_r = W_r / (T_{pr} + T_{qr},$$

then it is easy to see that (4.4) can be written as

$$\frac{\alpha T_{pr}}{\hat{x}_r} = T_{qr}. \qquad (4.5)$$

Let b_l denote the equilibrium queue length at link l. Then, the equilibrium
queueing delay at link l is b_l / c_l and thus,

$$T_{qr} = \sum_{l:l \in r} \frac{b_l}{c_l}.$$

Note that, if $\hat{y}_l < c_l$, then $b_l = 0$, where we recall that $\hat{y}_l$ is the equilibrium arrival rate at link l and is given by

$$\hat{y}_l = \sum_{r:l \in r} \hat{x}_l.$$

Further, notice that $\hat{y}_l \leq c_l$, since otherwise the equilibrium queue length will become ∞ and the system will not be in equilibrium. In summary, if the TCP-Vegas sources reach an equilibrium, then

$$\frac{\alpha T_{pr}}{x_r^*} = \sum_{l:l \in r} \frac{b_l}{c_l},$$

with

$$\frac{b_l}{c_l}(\hat{y}_l - c_l) = 0.$$

But these are precisely the Karush-Kuhn-Tucker conditions for optimality for the problem given in (4.3) with p_l, the Lagrange multiplier corresponding to the capacity constraint on link l, given by $p_l = b_l/c_l$. Thus, the theorem is proved. $\square$

If each source r uses a different parameter α_r as the control parameter for its Vegas algorithm, then it is easy to see that the equilibrium rates will maximize $\sum_r \alpha_r T_{pr} \log x_r$. In general, the TCP-Vegas algorithm attempts to allocate resources in a weighted proportionally-fair manner. We have interpreted the TCP-Vegas's equilibrium behavior (if an equilibrium is reached) in terms of the resource allocation problems posed in Chapter 2. However, it is difficult to directly interpret the dynamics of TCP-Vegas in terms of the congestion control algorithms presented in Chapter 3. For a relative comparison of the throughput that TCP-Vegas and TCP-Reno sources achieve in a network, we refer the interested reader to [77].

4.4 Random Early Detection (RED)

RED was introduced as a mechanism to break synchronization among TCP flows [27]. Currently, it is primarily used as a mechanism to maintain small queue lengths in the Internet. Under RED, a packet is dropped or marked with a certain probability which depends on the queue length. Instead of using the current queue length, RED maintains an exponentially-averaged estimate of the queue length and uses this to determine the marking probability. The basic idea is that, if the average queue length is large, then packets should be marked with a high probability to let the source know that the level of

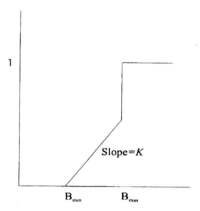

Fig. 4.1. The RED probability profile, with the average queue length on the x-axis and the marking or dropping probability on the y-axis.

congestion at the link is high. On the other hand, if the average queue length is small, the marking probability is low. An algorithm that sends congestion information from the link using the queue length is called an *active queue management* (AQM) algorithm. RED is an example of an AQM scheme.

Let b_{qv} be the average queue length at a link. Then, the marking probability at the link is determined according to the following profile:

$$f(b_{av}) = \begin{cases} 0, & \text{if } b_{av} \leq B_{min} \\ K(b_{av} - B_{min}), & \text{if } B_{min} < b_{av} \leq B_{max} , \\ 1, & \text{if } b_{av} > B_{max} \end{cases} \qquad (4.6)$$

where K is some constant, and B_{min} and B_{max} are some thresholds such that the marking probability is equal to 0 if the average queue length is below B_{min} and is equal to 1 if the average queue length is above B_{max}. The RED marking or dropping probability profile is shown in Figure 4.1. The averaged queue length is obtained by a weighting process as follows:

$$b_{av}(t+1) = \left(1 - \frac{1}{\varepsilon_l}\right) b_{av} + \frac{1}{\varepsilon_l} b(t),$$

where $b(t)$ is the queue length at time t. The continuous-time approximation of this equation is

$$\varepsilon \dot{b}_{av} = b_{av} + b.$$

The queue dynamics are given by

$$\dot{b} = (y - c)_b^+ , \qquad (4.7)$$

where y is the arrival rate at the link and c is the link bandwidth.

If we ignore the queue-length averaging and set $B_{min} = 0$, and consider only the linear portion of the RED marking profile, then the marking probability is given by $p = Kb$. Thus,

$$\dot{p} = K\dot{b} = K\left(y - c\right)_b^+,$$

which is nothing but the the router's algorithm to compute its Lagrange multiplier in the dual and primal-dual algorithms in Chapter 3.

4.5 Explicit Congestion Notification (ECN)

In today's Internet, congestion is indicated by dropping packets when links buffers are full, or by dropping packets according to an AQM scheme such as RED. However, to eliminate packet losses and the resulting inefficiency due to the retransmissions of such packets, a protocol called ECN has been proposed for congestion indication at the links [25]. Under the ECN mechanism, each packet has a bit which is normally set equal to zero. When a link detects incipient congestion (for example, by using RED), then this bit is set equal to one. When the destination receives a packet with the ECN bit set equal to one, it conveys this information back to the source in the ack packet. The TCP congestion control algorithm at the source can then treat this information in a manner similar to packet loss and reduce its congestion window. By providing early congestion notification, the ECN protocol can significantly reduce queueing delays and packet loss rates.

4.6 High-throughput TCP

Recall the increase portion of the congestion avoidance phase of Jacobson's algorithm: for every received ack, the source increments the window size by $1/cwnd$, where $cwnd$ is the current window size. Suppose that a source is transmitting data to a receiver at rate 1 Gbps and the RTT is 100 msecs. Assuming that a packet contains 10,000 bits, this corresponds to a window size of $10^9 * 0.1/10000$ or 10,000 packets. Next, suppose that a packet loss occurs when the window size is 10,000 packets. Then, the window is reduced to 5,000 packets and for every received ack, the window will be incremented by 1. Thus, it will take at least 5,000 RTTs or approximately at least 500 seconds for the window to build back up to the original value of 10,000 packets. Suppose that a source is using a fiber connection that is not shared by anyone else for this file transfer; even then a very rare packet loss due to transmission errors can cause a window decrease, followed by a very slow increase back to the original high-speed operation. This suggests that incrementing the window size by a constant amount for each received ack may be more suitable for high-speed operation.

Consider the following version of proportionally-fair controller:

$$\dot{x} = \kappa x(t - T)(w - x(t)q(t)), \tag{4.8}$$

where T is the round-trip time. Letting W denote the window size and using the approximate relationship $x = W/T$, where T is the RTT (which we assume does not change significantly), we have

$$\dot{W} = \kappa W(t - T)(w - \frac{W(t)}{T}q(t)).$$

(4.9)

Choosing $w = 1/T$, and discretizing the above equation over a small time interval of length δ gives

$$W(t + \delta) - W(t) = \kappa \frac{W(t - T)\delta}{T} - \kappa W(t) \left(\frac{W(t - T)q\delta}{T} \right).$$

(4.10)

The term $\frac{W(t-T)q\delta}{T}$ is the number of notifications of lost packets received at the source in the small interval $[t, t + \delta]$. Note that, since there is a feedback delay, this information is delayed by T time units. Thus, the decrease term in the above equation can be interpreted as decreasing the current window by an amount equal to κW for each dropped packet. Since $\frac{W(t-T)\delta}{T}$ is the number of acks or nacks (negative acknowledgments or dropped packet notifications) in a small interval of length δ, the increase term in (4.10) can be interpreted as a constant increase of κ for each ack or nack received at the source. Thus, this congestion control mechanism is more suitable for high-speed data transfers. We will refer to this algorithm, introduced in [94], as *High-throughput TCP*. A version of this was implemented by Tom Kelly under the name *Scalable TCP* and the details can be found in [53]. There are also other solutions to the high-speed transfer problem, and the interested reader is referred to [26] for an implementation called *High-Speed TCP* which, like (4.8) can be interpreted as a variant of the primal congestion control algorithm, and to [41] for a dual-based algorithm called *FAST*.

5

Linear Analysis with Delay: The single link case

In the previous chapters, we have studied the relationship between congestion control and resource allocation, and studied the stability of congestion control algorithms in the absence of feedback delay. We will now study the stability of congestion control algorithms in the presence of delay. In this chapter, we restrict our attention to the case of a single link accessed by one or many sources to get acquainted with the tools and techniques used to study stability. In the next chapter, we will extend these results to the case of general topology networks, accessed by a set of heterogeneous sources.

5.1 Single TCP-Reno source with droptail

Recall the source rate evolution of TCP-Reno:

$$\dot{x} = \kappa x(t-T) \left(\frac{1 - p(t-T)}{x(t)} - \beta x(t) p(t-T) \right), \tag{5.1}$$

where we take $p(t)$ to be the drop probability. Recall the notation introduced in the previous chapters, whereby we use $p(t)$ to denote the congestion feedback from a link and $q(t)$ to denote the congestion feedback from a route. However, since we are considering only a single link case in this chapter, the link feedback is also the same as the route feedback. Thus, we will only use $p(t)$ to denote the feedback from the network in this chapter. We assume that the link serves packets at the rate of c packets-per-second and has a buffer of size B packets. We assume that the queueing process at the link can be modelled as an $M/M/1/B$ queue with an arrival rate $x(t)$ and service rate c. Let $f(x)$ denote the drop probability in this queue; thus. $p(t) = f(x(t))$. The function $f(x)$ is given by [10]

$$f(x) = \frac{1 - (x/c)}{1 - (x/c)^{B+1}} \left(\frac{x}{c}\right)^B.$$

For TCP congestion control to work, packet losses must necessarily occur. Therefore, we make the assumption that the queue is nearly full most of the time. Thus, we make the approximation that the RTT, denoted by T, is a constant and is equal to the sum of the queueing delay B/c and the propagation delay, denoted by τ_p.

The goal of this section is to derive conditions on κ such that the congestion controller is asymptotically stable. In general, deriving conditions for global asymptotic stability is hard. Therefore, we will linearize the system around its equilibrium point and obtain conditions for local asymptotic stability. To this end, we use the notation $\hat{\ }$ to denote equilibrium quantities, for example, $\hat{x}$ denotes the equilibrium transmission rate of the source. We also use the notation δx to denote $x - \hat{x}$. Substituting $x = \hat{x} + \delta x$ in (5.1) and ignoring $(\delta x)^2$ and higher-order terms, we get

$$\delta\dot{x} = \kappa\hat{x}\left(-\frac{1-\hat{q}}{\hat{x}^2}\delta x - \frac{1}{\hat{x}}\delta q(t-T) + \beta\hat{x}\right). \tag{5.2}$$

Linearizing the relationship $p(t) = f(x(t))$ in a similar manner gives

$$\delta p(t) = f'(\hat{x})\delta x(t). \tag{5.3}$$

Taking the Laplace transform of (5.2) and (5.3) yields

$$\left(s + \frac{\kappa(1-\hat{p})}{\hat{x}} + \kappa\beta\hat{p}\hat{x}\right)x(s) + \kappa e^{-sT}(1 + \beta\hat{x}^2)p(s) = x_0,$$

and

$$p(s) = \hat{p}'x(s),$$

where x_0 is the value of $x(t)$ at time $t = 0$, $\hat{p} = f(\hat{x})$, and

$$\frac{1-\hat{p}}{\hat{p}} = \beta\hat{x}^2.$$

Note that we are abusing notation and using $x(\cdot)$ to denote the source rate as a function of time as well as to denote its Laplace transform. From the context, the usage should be clear. Thus,

$$(1 + G(s))x(s) = x_0,$$

where

$$G(s) = \frac{\kappa e^{-sT}\hat{p}'}{\hat{p}(s + \frac{\kappa(1-\hat{p})}{\hat{x}} + \kappa\beta\hat{p}\hat{x})}.$$

We want to apply the Nyquist criterion to $G(j\omega)$, where

$$G(j\omega) = \frac{\kappa T\hat{p}'}{\hat{p}}\frac{e^{-j\omega T}}{j\omega T + (\kappa\frac{1-\hat{p}}{\hat{x}} + \kappa\beta\hat{p}\hat{x})T}.$$

Now, we need the following lemma.

Lemma 5.1. *As ω is varied from $-\infty$ to ∞, $h_r(\omega) := \frac{e^{-j\omega T_r}}{\alpha_r + j\omega T_r}$ does not encircle the point $-2/\pi$.*

Proof.

$$h_r(\omega) = \frac{e^{-j\omega T_r}}{\alpha_r + j\omega T_r} = \frac{e^{-j(\omega T_r + \theta)}}{\sqrt{\alpha_r^2 + (\omega T_r)^2}}; \tag{5.4}$$

where

$$\theta_r = \text{Phase}(\alpha_r + j\omega T_r).$$

For the imaginary part of (5.4) to be equal to zero, the following condition must be satisfied:

$$\omega T_r + \theta_r = \pm n\pi, \qquad n = 0, 1, , 2, \ldots. \tag{5.5}$$

Since $\alpha_r > 0$, we have $|\theta_r| < \pi/2$. The real part of (5.4) is given by

$$\Re\left(h_r(\omega)\right) = \frac{\cos(\omega T_r + \theta_r)}{\sqrt{\alpha_r^2 + (\omega T_r)^2}}.$$

When ωT_r satisfies (5.5), the real part of $h_r(\omega)$ is positive when n is either zero or an even number. When n is an odd number, since $|\theta_r| < \pi/2$, $|\omega T_r| = |\pm n\pi - \theta_r| > \pi/2$. Therefore,

$$\Re\left(h_r(\omega)\right) = \frac{-1}{\sqrt{\alpha_r^2 + (\omega T_r)^2}} > -\frac{\pi}{2},$$

thus completing the proof. A plot of $\frac{e^{-j\omega T_r}}{j\omega T_r + \alpha_r}$ for $T_r = 1$, $\alpha_r = 2$ is shown in Figure 5.1, and a plot with $\alpha_r = 0.5$ is shown in Figure 5.2.

$\square$

From the above lemma, it is clear that the system is stable if $\kappa T \frac{\hat{p}'}{\hat{p}} < \pi/2$. For TCP, $\kappa = 1/T^2$. Thus, the stability condition becomes

$$\frac{\hat{p}'}{\hat{p}} < \frac{\pi T}{2}.$$

5.2 Multiple TCP sources with identical RTTs

In this case, the rate evolution of the ith TCP-Reno source is given by

$$\dot{x}_i = \kappa x_i(t - T)\left(\frac{1 - p(t - T)}{x_i(t)} + \beta x_i(t)p(t - T)\right), \tag{5.6}$$

where $p(t) = f(x(t))$ and, as in the previous section, $f(x(t))$ is the drop probability in an $M/M/1/B$ queue with arrival rate $x(t) = \sum_{j=1}^{N} x_i(t)$.

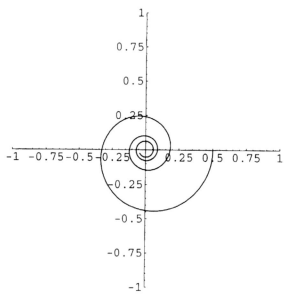

Fig. 5.1. Plot of $\frac{e^{-j\omega}}{j\omega+2}$ as ω is increased from 0 to ∞

Linearizing (5.6) as in the previous section and taking the Laplace transform of the resulting equation, it is easy to show that the linearized form of the above system is stable if the roots of the equation

$$s + \alpha_1 + \alpha_2 e^{-sT} = 0 \tag{5.7}$$

are in the left-half of the complex plane, where

$$\alpha_1 := \kappa \left(\frac{N(1-\hat{p})}{\hat{x}} + \beta\hat{p}\frac{\hat{x}}{N} \right),$$

$$\alpha_2 := \frac{N\kappa\hat{p}'}{\hat{p}},$$

$\hat{x}$ is the equilibrium value of $\sum_j x_j$, $\hat{p} = f(\hat{x})$ and $\hat{p}' = f(\hat{x}')$. Note that $\hat{x} = Ny$, where y is the equilibrium rate of a single user which is determined from the equation

$$\frac{1-\hat{p}}{y} = \beta y f(Ny).$$

As in the previous section, we can show that a sufficient condition for stability is given by

$$\frac{\hat{p}'N}{\hat{p}} < \frac{\pi T}{2}. \tag{5.8}$$

To examine how tight this condition is, we state the following lemma (see [35, 11] or [34, Theorem A.5]).

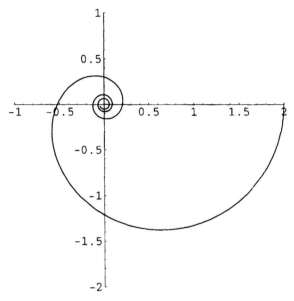

Fig. 5.2. Plot of $\frac{e^{-j\omega}}{j\omega+0.5}$ as ω is increased from 0 to ∞

Lemma 5.2. *The roots of (5.7) lie in the left-half plane if and only if one of the following conditions is satisfied:*

1. $\alpha_1 \geq \alpha_2$,
2. $\alpha_1 < \alpha_2$ and

$$\alpha_2 T\sqrt{1 - \frac{\alpha_1^2}{\alpha_2^2}} < \arccos\left(-\frac{\alpha_1}{\alpha_2}\right).$$

$\square$

Consider a link of capacity 1 Gbps shared by 1000 users. We assume that each packet contains 8000 bits. Thus, the link capacity c can be expressed as $125,000$ packets-per-sec. To see the effect of RTT on the stability properties of TCP, we consider the following two examples on this link.

Example 5.3. In this experiment, we fix the buffer size B to be 125 packets for a maximum queueing delay of 1 msec. Noting that $\kappa\beta = 0.5$ for TCP, the system is stable for $\tau_p \leq 25.3$ msecs., and it is unstable for larger τ_p. The sufficient condition (5.8) for stability fails for $\tau \geq 19.8$ msecs. $\square$

Example 5.4. In this experiment, we choose the buffer size to be equal to $\lfloor C * \tau_p \rfloor$ so that the maximum queueing delay is equal to τ_p. The system is stable for $\tau_p < 12.7$ msecs., and is unstable for larger τ_p msecs. The sufficient condition (5.8) for stability fails for $\tau \geq 9.8$ msecs. $\square$

We conclude the following from the previous two examples:

- While the Nyquist criterion does not give a tight condition for stability, the sufficient condition (5.8) cannot be improved upon significantly.
- TCP is unstable for large delays and thus, it is important to design congestion controllers that are stable for large RTTs. We will address this problem of designing stable, scalable congestion controllers in the next chapter.

The number of users of the Internet is ever-increasing, and so are the capacity of the links in the Internet. Further, as the link speeds are increased, one can keep the delay in the link constant, by proportionally increasing the buffer size. Thus, it is reasonable to suppose that as the number of users increase, the link capacity and buffer size increase proportionally. Thus, we can obtain a reasonable approximation for the $M/M/1/B$ drop probability for large networks by letting B go to ∞ in the expression

$$\frac{1 - (x/c)}{1 - (x/c)^{B+1}} (x/c)^B.$$

It is easy to see that

$$\lim_{B \to \infty} \frac{1 - (x/c)}{1 - (x/c)^{B+1}} (x/c)^B = \begin{cases} 0, & \text{if } x/c < 1, \\ 1 - \frac{c}{x}, & \text{if } x/c \geq 1. \end{cases}$$

In other words, a reasonable approximation for the $M/M/1/B$ drop probability, $f(x)$, in a link with capacity c is given by [62, 58]:

$$f(x) = \left(1 - \frac{c}{x}\right)^+.$$

Note that this approximation has the following simple interpretation: when the arrival rate is less than the link capacity, then the number of dropped packets is zero. On the other hand, when the arrival rate is greater than or equal to the link capacity, then the rate at which packets are dropped is equal to the arrival rate minus the service rate. Using this approximation, it is easy to see that the conclusions of Examples 1 and 2 continue to hold, i.e., TCP is unstable for large RTTs.

5.3 TCP-Reno and RED

Next, we derive conditions for the stability of a TCP source accessing a link where the feedback is provided by the RED algorithm introduced in the previous chapter.

$$\dot{x} = \kappa x(t - T) \left(\frac{1 - p(t - T)}{x(t)} - \beta x(t) p(t - T) \right). \tag{5.9}$$

Linearizing the above equation, we obtain

$$\delta \dot{x} = -\kappa \hat{x} \left(\frac{1 - \hat{p}}{\hat{x}^2} \delta x(t) + \frac{1}{\hat{x}} \delta p(t - T) + \beta \hat{x} \delta p(t - T) - \beta \hat{p} \delta x(t) \right).$$

Taking Laplace transforms, we get

$$\left(s + \frac{\kappa \hat{x}(1 - \hat{p})}{\hat{x}^2} + \kappa \beta \hat{p} \right) x(s) - x_0 = -\kappa e^{-sT} p(s) - \beta \kappa \hat{x}^2 e^{-sT} p(s),$$

where x_0 is the initial condition of the source, i.e., the source rate at time 0. Since

$$p(t) = f(b_{av}(t)),$$

after linearization and taking the Laplace transform, we get

$$p(s) = \hat{p}' b_{av}(s),$$

where $\hat{p}' = f'(\hat{x})$. Further, from the dynamics of the averaged queue length (4.7), we get

$$(1 + \varepsilon s) b_{av}(s) = b(s) + b_{av0},$$

where b_{av0} is the initial condition for the average queue length $b_{av}(t)$. From the queue dynamics, we get

$$sb(s) - b_0 = x(s).$$

Thus, the transfer function from the initial conditions to $x(s)$ can be written as

$$\left(s + \kappa \frac{(1 - \hat{p})}{\hat{x}} + \kappa \beta \hat{p} + e^{-sT} \frac{\kappa \hat{p}(1 + \beta \hat{x}^2)}{(1 + \varepsilon s)s} \right) x(s) = u(s),$$

where

$$u(s) = x_0 + \frac{e^{-sT} \kappa \hat{p}(1 + \beta \hat{x}^2)}{s(1 + \varepsilon s)} (b_0 + s b_{av0}).$$

Thus,

$$x(s) = \frac{s(1 + \varepsilon s) x_0 + e^{-sT} \kappa \hat{p}(1 + \beta \hat{x}^2)(b_0 + s b_{av0})}{s(1 + \varepsilon s)(s + \frac{\kappa(1 - \hat{p})}{\hat{x}} + \kappa \beta \hat{p}) + e^{-sT} \kappa \hat{p}(1 + \beta \hat{x}^2)}. \tag{5.10}$$

For the Fourier transform inverse of the above equation to go to zero as $t \to \infty$, the solutions to s obtained by setting the denominator in the expression for $x(s)$ to zero should all lie in the complex left-half plane. It is straightforward to verify that

$$s = 0, \qquad s = -1/\varepsilon, \qquad s = -\kappa \frac{(1 - \hat{p})}{\hat{x}} - \kappa \beta \hat{p}$$

cannot be solutions to the equation

$$(1 + \varepsilon s)s(s + \kappa \frac{(1 - \hat{p})}{\hat{x}} + \kappa \beta \hat{p}) + e^{-sT} \kappa \hat{p}(1 + \beta \hat{x}^2) = 0.$$

Thus, the denominator in the expression for $x(s)$ in (5.10) being equal to zero, is equivalent to

$$1 + G(s) = 0,$$

where

$$G(s) = e^{-sT} \frac{\kappa \hat{p}(1 + \beta \hat{x}^2)}{(1 + \varepsilon s)s(s + \kappa \frac{(1-\hat{p})}{\hat{x}} + \kappa \beta \hat{p})}.$$

Therefore, we can apply the Nyquist criterion to $G(s)$. Let

$$G(j\omega) = r(\omega)e^{-j\theta(\omega)}$$

where

$$r = \frac{\kappa \hat{p}(1 + \beta \hat{x}^2)}{|\omega|\sqrt{1 + \varepsilon^2 \omega^2}\sqrt{\omega^2 + \kappa^2 \left(\frac{1-\hat{p}}{\hat{x}} + \beta \hat{p}\right)^2}},$$

and

$$\theta = \omega T + \frac{\pi}{2} + \underbrace{\arctan(\varepsilon \omega)}_{=:a} + \underbrace{\arctan\left(\frac{\omega}{\kappa(\frac{1-\hat{p}}{\hat{x}} + \beta \hat{p})}\right)}_{=:b}.$$

For the imaginary part of $G(j\omega)$ to be equal to zero, we need

$$\arctan(b) = \pm n\pi - \omega T - \frac{\pi}{2} - \arctan(b),$$

or, equivalently,

$$\varepsilon \omega = \cot\left(\omega T + \arctan\left(\frac{\omega T}{\kappa(\frac{1-\hat{p}}{\hat{x}} + \beta \hat{p})}\right)\right). \tag{5.11}$$

Note that $\omega = 0$ cannot be a solution to the above equation. Let ω^* be the smallest positive solution to (5.11). A sufficient condition to ensure that the plot of $G(j\omega)$ does not encircle the point -1, is to make sure that the real part of $G(j\omega)$ is greater than -1 whenever the imaginary part of $G(j\omega)$ is equal to zero. Thus, a sufficient condition for stability is $|r| < 1$, which can be ensured if

$$\frac{\kappa \hat{p}(1 + \beta \hat{x}^2)}{(\omega^*)^2} < 1.$$

In a similar manner, a stability condition for the case of multiple sources can be derived. It has been shown in [76] that TCP-Reno with RED is also stable only for small RTTs, and the system becomes unstable for large RTTs. Further, the averaging parameter ε also plays a crucial role in the stability of the system, and thus, has to be designed carefully. A congestion indication mechanism called the proportional-integral (PI) controller which has better stability properties than RED has been proposed in [37].

5.4 Proportionally-fair controller

Consider the delayed feedback version of the primal proportionally-fair controller introduced in Chapter 3:

$$\dot{x} = \kappa \left(w - x(t-T)f(x(t-T)) \right). \tag{5.12}$$

From now on, for convenience, we will refer to the controller (5.12), with $\kappa(x) \equiv \kappa \; \forall x$ for some constant κ, as the proportionally-fair controller. Linearizing around the equilibrium point, we get

$$\delta\dot{x} = -\kappa(\hat{p} + \hat{x}\hat{p}')\delta x(t-T),$$

where $\hat{p}' = f'(\hat{x})$. Taking Laplace transforms, the above differential equation becomes

$$\left(s + \kappa(\hat{p} + \hat{x}\hat{p}')e^{-sT} \right) x(s) = \delta x_0.$$

Thus, if the roots of the characteristic equation

$$s + \kappa(\hat{p} + \hat{x}\hat{p}')e^{-sT} = 0$$

lie in the complex left-half plane, then the system is stable. Assuming that $\hat{p} + \hat{x}\hat{p}' \neq 0$, it is easy to see that $s = 0$ cannot be a solution to the characteristic equation. Thus, the characteristic equation can be rewritten as

$$1 + \kappa T(\hat{p} + \hat{x}\hat{p}')\frac{e^{-sT}}{sT} = 0.$$

Define

$$G(s) = \kappa T(\hat{p} + \hat{x}\hat{p}')\frac{e^{-sT}}{sT}.$$

Applying the Nyquist criterion, we are interested in verifying that the plot of $G(j\omega)$ does not encircle the point $-1 + j0$ as ω is varied from $-\infty$ to $+\infty$. We first present the following lemma.

Lemma 5.5. *As ω is varied from $-\infty$ to ∞, $\frac{\pi}{2}\frac{e^{-j\omega T_r}}{j\omega T_r}$ does not encircle the point -1.*

Proof. First note that

$$\frac{e^{-j\omega T_r}}{j\omega T_r} = \frac{e^{-j(\omega T_r + \pi/2)}}{\omega T_r}. \tag{5.13}$$

For the imaginary part of (5.13) to be equal to zero, ωT_r must be equal to $(2n+1)\pi/2$, for $n = 0, \pm 1, \pm 2, \dots$. Thus, $|\omega T_r| \geq |\pi/2|$ when the imaginary part of (5.13) is equal to zero. When $|\omega T_r| \geq |\pi/2|$, the real part of (5.13), given by

$$\frac{\cos(\omega T_r + \pi/2)}{\omega T_r} = \frac{-\sin \omega T_r}{\omega T_r},$$

is easily seen to be greater than or equal to $-2/\pi$, thus completing the proof. A plot of $\frac{e^{-j\omega T_r}}{j\omega T_r}$ for $T_r = 1$ is shown in Figure 5.3. $\qquad\square$

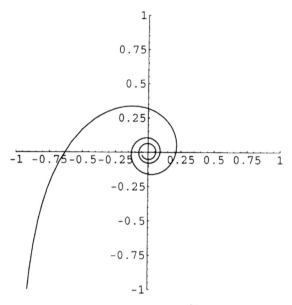

Fig. 5.3. Plot of $\frac{e^{-j\omega}}{j\omega}$ as ω is increased from 0 to ∞

Therefore, if

$$\kappa(\hat{p} + \hat{x}\hat{p}') < \frac{\pi}{2},$$

then $G(j\omega)$ does not encircle $-1 + j0$ and the linearized system is stable.

5.5 High-throughput TCP

Consider the following congestion control algorithm introduced in Section 4.6:

$$\dot{x} = \kappa x(t - T)\left(w - x(t)f(x(t - T))\right). \tag{5.14}$$

Note that this congestion-control algorithm also achieves proportional fairness when all sources in a network use such a controller. However, to distinguish between this controller and (5.12), we call this the high-throughput TCP, and (5.12), the proportionally-fair controller.

Linearizing (5.14) around the equilibrium yields

$$\delta\dot{x} = -\kappa\hat{x}\left(\hat{p}\delta x + \hat{x}\hat{p}'\delta x(t - T)\right).$$

In the Laplace domain, the above delay-differential equation becomes

$$\left(s + \kappa\hat{x}\hat{p} + \kappa\hat{x}^2\hat{p}'e^{-sT}\right)x(s) = x_0.$$

Thus, the characteristic equation is given by

$$s + \kappa \hat{x} \hat{p} + \kappa \hat{x}^2 \hat{p}' e^{-sT} = 0.$$

We are interested in conditions under which the characteristic equation has all its roots in the complex left-half plane. Noting that $s == -\kappa \hat{x} \hat{p}$ cannot be a root of the characteristic equation, we can rewrite the characteristic equation as

$$1 + \frac{2}{\pi} \kappa \hat{x}^2 \hat{p}' T \frac{\pi}{2} \frac{e^{-sT}}{sT + \kappa \hat{x} \hat{p} T} = 0.$$

From Lemma 5.1, we know that

$$\frac{\pi}{2} \frac{e^{-j\omega T}}{j\omega T + \kappa \hat{x} \hat{p} T}$$

does not encircle the point $-1 + j0$ as ω is varied from $-\infty$ to $+\infty$. Thus, the linear system is stable if

$$\kappa \hat{x}^2 \hat{p}' T < \frac{\pi}{2}.$$

Noting that

$$w = \hat{x} \hat{p},$$

it is useful to rewrite the stability condition as

$$\kappa w T \frac{\hat{x} \hat{p}}{\hat{p}'} < \frac{\pi}{2}.$$

Thus, the stability condition can be decomposed into a condition on the source control gain κ,

$$\kappa w T < \frac{\pi}{2B},$$

and a condition on the marking function at the router

$$\frac{\hat{x} \hat{p}}{\hat{p}'} \leq B$$

where $B > 0$. It is easy to see that the router condition is satisfied if, for example,

$$p(x) = \left(\frac{x}{c}\right)^B,$$

which is the probability that the queue length exceeds B in an $M/M/1$ queue with arrival rate x and capacity c.

5.6 Dual algorithm

The source end of the dual algorithm with feedback delay is given by

$$U'(x) = p(t - T),$$

and the router end is given by

$$\dot{p} = \alpha(x - c)_p^+,$$

for some $\alpha > 0$. If we choose $\alpha = 1/c$, then the price dynamics at the router becomes

$$\dot{p} = \frac{1}{c}(x - c)_p^+.$$

Now the price can be interpreted as the delay at the link since the delay d of a packet which sees a buffer level b upon arrival will be equal to b/c. Thus,

$$\dot{d} = \frac{1}{c}\dot{b} = \frac{1}{c}(x - c)_b^+,$$

which is also the dynamics of the link price. Linearizing the source and price dynamics around the equilibrium gives

$$U''(\hat{x})\delta x(t) = \delta p(t - T)$$

and

$$\delta \dot{p} = \frac{1}{c}\delta x,$$

where $\hat{x} = c$. Taking Laplace transforms, it is easy to see that the characteristic equation is given by

$$1 + \frac{T}{c|U''(\hat{x})|}\frac{e^{-sT}}{sT} = 0.$$

Applying the Nyquist criterion and using Lemma 5.5, it is easy to see that a sufficient condition for stability is given by

$$\frac{T}{c|U''(\hat{x})|} < 1.$$

If $U(x) = \log x$, then this condition becomes

$$cT < 1.$$

The stability condition is satisfied only for small values of the bandwidth-delay product cT. On the other hand if $U(x)$ is also a function of T, then the condition may be satisfied for large cT. Thus, the dual solution is scalable for large values of the bandwidth-delay product only if the utility function includes the RTT. As we will see in the next chapter, this solution is not fair when a network is shared by many sources with different RTTs. However, there is also a way to fix this problem and this will also be discussed in the next chapter.

5.7 Primal-dual algorithm

Consider the primal-dual controller, with the primal algorithm at the source given by

$$\dot{x} = \kappa x(t - T)\,(w - x(t)p(t - T))$$

and the dual algorithm at the router given by

$$\dot{p} = g(p)(x - c)_p^0,$$

for some $g(p) > 0$. Linearizing around the equilibrium gives

$$\delta\dot{x} = \kappa\hat{x}(\hat{x}\delta p(t - T) + \hat{p}\delta x)$$

and

$$\delta\dot{p} = g(\hat{p})x.$$

Taking Laplace transforms, we get

$$(s + \kappa\hat{x}^2)x(s) - \delta x_0 = -\kappa\hat{x}^2 e^{-sT}p(s)$$

and

$$sp(s) - \delta p_0 = g(\hat{p})x(s).$$

Thus, the characteristic equation is given by

$$1 + g(\hat{p})T\frac{\kappa\hat{x}^2 T e^{-sT}}{sT(sT + \kappa\hat{x}^2 T)}.$$

We now state the following lemma.

Lemma 5.6. *As x is varied from $-\infty$ to ∞, $f(x) := \dfrac{\theta e^{-jx}}{jx(jx + \theta)}$ crosses the real line at a point to the right of -1 for all $\theta > 0$.*

Proof. Suppose $f(x)$ crosses the real line at $-b$ when $x = x_0$. Then we just need to prove that $b < 1$. We have:

$$\tan x_0 = \frac{\theta}{x_0}.$$

So

$$x_0 \tan x_0 = \theta,$$

$$b = \frac{\theta}{x_0\sqrt{x_0^2 + \theta^2}} = \frac{\cos x_0}{x_0^2}x_0\tan x_0 = \frac{\sin x_0}{x_0} < 1.$$

A plot of $\frac{\theta e^{-jx}}{jx(jx+\theta)}$ is shown in Figure 5.4. $\square$

Thus, the primal-dual algorithm is stable if

$$g(\hat{p})T < 1.$$

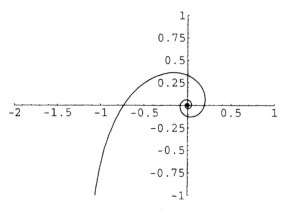

Fig. 5.4. Plot of $\frac{\theta e^{-jx}}{jx(jx+\theta)}$ for $\theta = 5.0$, as x is increased from 0 to ∞

5.8 Appendix: The Nyquist criterion

To check if a linear system is stable, one has to verify if the poles of its transfer function $L(s)$ lie in the complex left-half plane. In general, finding the roots of such an equation could be difficult. The Nyquist theorem provides an alternate condition to verify stability that is often easier to apply. The Nyquist criterion is given in the following theorem [28].

Theorem 5.7. *Let C be a closed contour in the complex plane such that no poles or zeros of $L(s)$ lie on C. Let Z and P denote the number of poles and zeros, respectively, of $L(s)$ that lie inside C. Then, the number of times that $L(j\omega)$ encircles the origin in the clockwise direction, as ω is varied from $-\infty$ to ∞, is equal to $Z - P$.*   □

Typically, one is interested in applying the above theorem to the transfer function $L(s)$ of a system. In other words, the input and output Laplace transforms of the system, say $u(s)$ and $y(s)$, respectively, are related by $y(s) = L(s)u(s)$. Such an input output relationship may be the result of a negative feedback as shown in Figure 5.5. In this case, it is easy to see that

$$\frac{y(s)}{u(s)} = \frac{G(s)}{1 + G(s)H(s)}.$$

Assuming that there are no pole-zero cancellations in the numerator and denominator, one needs to check if the roots of

$$1 + G(s)H(s) = 0$$

lie in the left-half plane to ensure stability. Thus, instead of checking that the poles of $L(s)$ lie in the left-half plane, we can check if the zeros of $1 + G(s)H(s)$ lie in the left-half plane. In terms of Theorem 5.7, we want $Z = 0$, where

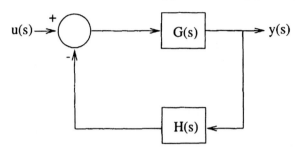

Fig. 5.5. A negative feedback interconnection

Z is the number of zeros of $1 + G(s)H(s)$ in the complex right-half plane. Suppose that $1 + G(s)H(s)$ has no poles in the right-half of the complex plane and the imaginary axis. Then, we can apply the Nyquist criterion, with the contour being the imaginary axis and fictitious half-circle connecting $+j\infty$ and $-j\infty$ that lies entirely on the right-half plane. Applying the Nyquist criterion, the stability of the system is then equivalent to saying that there are no encirclements of the origin by the plot of $1 + G(j\omega)H(j\omega)$ as ω is varied from $-\infty$ to ∞. Suppose that P, the number of poles of $1 + G(s)H(s)$, is greater than zero, then to ensure stability we need P counterclockwise encirclements of the origin by the plot of $1 + G(j\omega)H(j\omega)$ as ω is varied from $-\infty$. If there is a pole of $G(j\omega)H(j\omega)$ on the imaginary axis, then again the Nyquist stability criterion can be applied, but the contour has to be deformed by an infinitesimal amount around the pole; for details, see [28].

6

Linear Analysis with Delay: The network case

In this chapter, we extend the delay analysis of Chapter 5 to the case of a network. As in Chapter 5, we linearize the system around its equilibrium point and derive conditions for local stability. We consider three main classes of controllers:

1. We consider different variants of the primal controllers introduced in Chapter 3 with delays in both the forward and backward paths. We also consider the AVQ algorithm to achieve full utilization.

2. Next, we consider the dual controllers introduced in Chapter 3. Unlike the case with no delays, when delays are introduced into the network, we will see that, to ensure the stability of the dual controllers, only a certain class of source utility functions is allowed. However, borrowing the key idea from AVQ, we will see that a parameter in each source's utility function can be dynamically chosen so that the network equilibrium can be interpreted as maximizing the sum of arbitrary source utilities. Another difference between the dual and primal controllers is that the congestion feedback in the dual algorithm is the delay in the network, whereas in the primal algorithm, the congestion feedback is in the form of packet marks or losses.

3. Finally, we will consider primal-dual algorithms which allow one to model today's Internet, by allowing the source control algorithms to be like TCP-Reno while the congestion feedback is closely related to the RED algorithm.

6.1 Primal controllers

6.1.1 Proportionally-fair controller

Consider the congestion controller

$$\dot{x}_r = \kappa_r(w_r - x_r(t - T_r)q_r(t)), \tag{6.1}$$

where $q_i(t)$ is the delay path price given by

$$q_r(t) = \sum_{l \in r} p_l(t - d_2(l, r)), \qquad (6.2)$$

where $d_1(l, r)$ is the propagation delay from source r to link l, $d_2(l, r)$ is the propagation delay from link l to source r, and $T_r = d_1(l, r) + d_2(l, r)$ for any $l \in r$ is the round-trip delay for source r. In Internet parlance, round-trip delay is also referred to as round-trip time and abbreviated as RTT. Linearizing (6.1), we get

$$\delta \dot{x}_r = -\kappa_r \left(\hat{q}_r \delta x_r(t - T_r) + \hat{x}_r \delta q_r(t) \right). \qquad (6.3)$$

Taking the Laplace transform of the above equation and noting the fact that $\hat{q}_r = w_r / x_r$ yields

$$\left(s + \frac{\kappa_r w_r}{\hat{x}_r} e^{-sT_r} \right) x_r(s) = -\kappa_r \hat{x}_r q_r(s) + \delta x_r(0),$$

where we have abused the notation a little to denote the Laplace transform of $\delta x_r(t)$ by $x_r(s)$. Let $D(s)$ denote the diagonal matrix of RTTs in the Laplace domain, i.e., $D(s) := diag\{e^{-sT_r}\}$. Also, let $X = diag\{\hat{x}_r\}$, $W = diag\{w_r\}$ and $K = diag\{\kappa_r\}$. Then, the above Laplace transform equation can be written as

$$(sI + KWX^{-1}D(s))\mathbf{x}(s) + KX\mathbf{q}(s) = \mathbf{x}_0, \qquad (6.4)$$

where $\mathbf{x}_0$ is the vector of initial states. Linearizing and taking the Laplace transform of the path price equation (6.2) yields

$$q_r(s) = \sum_{l:l \in r} e^{-sd_2(l,r)} p_l(s) = e^{sT_r} \sum_{l:l \in r} e^{sd_1(l,r)} p_l(s).$$

Let $R(s)$ denote an $L \times R$ Laplace domain routing matrix that includes both routing and delay information, whose $(r, l)^{\text{th}}$ entry is defined as follows:

$$R_{lr} = \begin{cases} e^{-sd_1(l,r)}, & \text{if } l \in r, \\ 0, & \text{else.} \end{cases}$$

Thus,

$$\mathbf{q}(s) = D(s)R^T(-s)\mathbf{p}(s).$$

Suppose that $p_l(t)$ is generated according to a marking function $f_l(y_l(t))$; then

$$p_l(t) \approx f_l(\hat{y}_l) + \hat{p}'_l \delta y_l(t) = \hat{p}_l + f'_l(\hat{y}_l) \delta y_l(t),$$

where $\hat{p}'_l := f'_l(\hat{y}_l)$. Linearizing this relationship gives

$$\delta p_l(t) = \hat{p}'_l \delta y_l(t).$$

Thus, in the Laplace domain, we get

$$\mathbf{p}(s) = F_p \mathbf{y}(s),$$

where $F_p = diag\{\hat{p}'_l\}$. Now, noting that

$$y_l(t) = \sum_{r:l\in r} x_r(t - d_1(l,r))$$

gives

$$\mathbf{y}(s) = R(s)\mathbf{x}(s).$$

Thus,

$$\mathbf{q}(s) = D(s)R^T(-s)F_p R(s)\mathbf{x}(s). \tag{6.5}$$

Defining $\mathcal{K} = diag\{\kappa_r T_r\}$ and $D_T(s) = diag\{\frac{e^{-sT_r}}{sT_r}\}$, from (6.4) and (6.5), we get

$$s\left(I + \mathcal{K}WX^{-1}D(s) + \mathcal{K}X\mathcal{D}(s)R^T(-s)F_p R(s)\right)\mathbf{x}(s) = \mathbf{x}_0.$$

From basic control theory, the above system is stable if its poles lie in the left-half of the complex plane, i.e., the solutions to

$$det(sI + sG(s)) = 0$$

should have negative real parts, where

$$G(s) = \mathcal{K}WX^{-1}\mathcal{D}(s) + \mathcal{K}X\mathcal{D}(s)R^T(-s)F_p R(s). \tag{6.6}$$

It is easy to see that $s = 0$ cannot be a solution to $det(sI + sG(s)) = 0$. Therefore, we can equivalently check if the roots of $det(I + G(s)) = 0$ have negative real parts.

From the multivariable Nyquist criterion (see Appendix 6.4), the stability condition is equivalent to the following statement: the eigenvalues of $G(j\omega)$, denoted by $\sigma(G(j\omega))$, should not encircle the point -1.

In what follows, we will make use of the following facts:

1. If D_1 and D_2 are diagonal matrices, and A is any other matrix, then $D_1 D_2 A = D_2 D_1 A$.
2. If A and B are two square matrices, then the non-zero eigenvalues of AB are the same as the non-zero eigenvalues of BA.
3. If A is a positive-definite matrix, then

$$\lambda_{min}(A) \le \mathbf{x}^T A \mathbf{x} \le \lambda_{max}(A),$$

for any vector $\mathbf{x}$ such that $\mathbf{x}^T\mathbf{x} = 1$.
4. For an invertible matrix A, if λ is an eigenvalue of A, then $1/\lambda$ is an eigenvalue of A^{-1}.

5. The magnitude of any eigenvalue of a matrix A is upper-bounded by the maximum row sum, where the row sum refers to the sum of the absolute values of the elements of a row. In other words,

$$|\lambda(A)| \leq \max_i \sum_j |A_{ij}|,$$

where $\lambda(A)$ is any eigenvalue of A.

6. Let $\|A\|_\infty$ denote the maximum row sum. Then,

$$\|AB\|_\infty \leq \|A\|_\infty \|B\|_\infty.$$

From the above facts, it is easy to see that

$$\sigma(G(j\omega)) = \sigma\left(\mathcal{K}X(WX^{-2}D(j\omega) + D(j\omega)R^T(-j\omega)F_pR(j\omega))\right)$$
$$= \sigma\left(\mathcal{K}X(WX^{-2} + R^T(-j\omega)F_pR(j\omega))D(j\omega)\right)$$
$$= \sigma\left(\sqrt{\mathcal{K}X}(WX^{-2} + R^T(-j\omega)F_pR(j\omega))\sqrt{\mathcal{K}X}D(j\omega)\right).$$

Define

$$Q = \sqrt{\mathcal{K}X}(WX^{-2} + R^T(-j\omega)F_pR(j\omega))\sqrt{\mathcal{K}X}.$$

Note that Q is a positive definite matrix due to the fact that $\sqrt{\mathcal{K}X}WX^{-2}\sqrt{\mathcal{K}X}$ is positive definite and $\sqrt{\mathcal{K}X}R^T(-j\omega)F_pR(j\omega)\sqrt{\mathcal{K}X}$ is positive semi-definite. Thus, the eigenvalues of Q are real and positive. Let λ be an eigenvalue of Q and $\mathbf{v}$ be the corresponding eigenvector such that

$$\|\mathbf{v}\|^2 = \mathbf{v}^*\mathbf{v} = |v_1|^2 + |v_2|^2 + \cdots + |v_R|^2 = 1.$$

Then,

$$Q\mathcal{D}\mathbf{v} = \lambda\mathbf{v}$$

which implies that

$$\mathcal{D}\mathbf{v} = \lambda Q^{-1}\mathbf{v},$$

where we have used the fact that Q is positive definite to conclude that it is invertible. Thus,

$$\mathbf{v}^*\mathcal{D}\mathbf{v} = \lambda\mathbf{v}^*Q^{-1}\mathbf{v},$$

or,

$$\lambda = \frac{\mathbf{v}^*\mathcal{D}\mathbf{v}}{\mathbf{v}^*Q^{-1}\mathbf{v}}.$$

From the definition of $\mathcal{D}$,

$$\mathbf{v}^*\mathcal{D}\mathbf{v} = \sum_r |v_r|^2 \frac{e^{-j\omega T_r}}{j\omega T_r}.$$

From Lemma 5.13 in the previous chapter, it is easy to see that $\mathbf{v}^*\mathcal{D}\mathbf{v}$ does not encircle the point $-2/\pi$. Next, we note that

$$\mathbf{v}^* Q^{-1} \mathbf{v} \geq \lambda_{min}(Q^{-1}) = 1/\lambda_{max}(Q).$$

Thus,

$$\sigma(Q\mathcal{D}) \subset \lambda_{max}(Q) \sum_r |v_r|^2 \frac{e^{-j\omega T_r}}{j\omega T_r}.$$

Therefore, if $\lambda_{max(Q)} < \pi/2$, then $\sigma(Q\mathcal{D}(j\omega))$ does not encircle the point -1. Note that $\sigma(Q) = \sigma(\tilde{Q})$, where

$$\tilde{Q} := \lambda(KWX^{-1} + KXR^T(-j\omega)F_pR(j\omega).$$

Now, using the fact that the eigenvalues of $\tilde{Q}$ are upper-bounded by the maximum row sum, and noting that the sum of the absolute values of the rth row is given by $\kappa_r T_r(\hat{q}_r + \sum_{l:l \in r} f_l'(\hat{y}_l)\hat{y}_l)$, we get the following result [93].

Theorem 6.1. *If*

$$\kappa_r T_r\left(\hat{q}_r + \sum_{l:l \in r} f_l'(\hat{y}_l)\hat{y}_l\right) < \frac{\pi}{2}, \qquad \forall r,$$

then the congestion-control scheme (6.3) is locally, asymptotically stable.

Proof. See discussion prior to the theorem. □

Note the strikingly decentralized nature of the above result, each source needs information only about its route to choose its controller gain κ_r. In particular, if the marking probability on link l is chosen to be $f_l(y_l) = (y_l/\bar{c}_l)^B$, then $f_l'(y_l)y_l = Bf_l(y_l)$. Thus, the local stability condition becomes

$$\kappa_r T_r(1 + B)\hat{q}_r < \frac{\pi}{2},$$

which is only a function of the path price and a global parameter B. The expression also provides a tradeoff between choice of the marking function $f_l(y_l)$ and the controller gain κ_r. If the rate at which $f_l(y_l)$ increases is very large (i.e., B is large), then κ_r has to be small.

The result in the above theorem was conjectured by Johari and Tan in [43] and a slightly weaker version of the theorem was first proved in [71].

6.1.2 High-throughput TCP with rate-based feedback

Consider the congestion controller

$$\dot{x}_r = \kappa_r x_r(t - T_r)\left(w_r - x_r(t)q_r(t)\right), \tag{6.7}$$

where $q_r(t)$ remains the same as in the previous section. Linearizing and taking the Laplace transform yields

$$x_r(s) = \left(\frac{-\kappa_r T_r \hat{x}_r w_r}{(sT_r + \kappa_r T_r \hat{x}_r \hat{q}_r)\hat{q}_r} \right) q_r(s) + \delta x_r(0). \tag{6.8}$$

Equivalently, the transfer function from $\mathbf{q}$ to $\mathbf{x}$ can be written as

$$\mathbf{x}(s) = -M_1 M_2(s)\mathbf{q}(s) + \mathbf{x}_0,$$

where

$$M_1 = \mathcal{K}XW diag\{1/\hat{q}_r\}, \tag{6.9}$$

and

$$M_2(s) = diag\left\{ \frac{1}{sT_r + \alpha_r} \right\}, \tag{6.10}$$

and $\alpha_r := \kappa_r T_r \hat{x}_r \hat{q}_r$. The transfer function from $\mathbf{x}$ to $\mathbf{q}$ remains the same as in the previous section. Thus, the overall transfer function of the network becomes

$$\left(I + \tilde{\mathcal{D}}(s)M_1 R^T(-s)F_p R(s) \right) \mathbf{x}(s) = \mathbf{x}_0,$$

where

$$\tilde{\mathcal{D}}(s) = diag\left\{ \frac{e^{-sT_r}}{sT_r + \alpha_r} \right\}. \tag{6.11}$$

Since each entry of M_1 is positive, it is easy to see that

$$\sigma(M_1 R^T(-j\omega)F_p R(j\omega)) = \sigma(\sqrt{M_1} R^T(-j\omega)F_p R(j\omega)\sqrt{M_1}).$$

Further, since $R^T(-j\omega) = R^*(j\omega)$, where A^* denotes the conjugate transpose of A, $\sqrt{M_1} R^T(-j\omega)F_p R(j\omega)\sqrt{M_1}$ is a positive semi-definite matrix. Thus, its eigenvalues are non-negative. Now,

$$\lambda(M_1 R^T(-j\omega)F_p R(j\omega))$$

$$= \lambda(\mathcal{K}W diag\left\{ \frac{1}{\hat{q}_r} \right\} R^T(-j\omega)F_p diag\{\hat{y}_l\} diag\left\{ \frac{1}{\hat{y}_l} \right\} R(j\omega)X)$$

$$\leq \|\mathcal{K}W diag\left\{ \frac{1}{\hat{q}_r} \right\} R^T(-j\omega)F_p diag\{\hat{y}_l\}\|_\infty \cdot \|diag\left\{ \frac{1}{\hat{y}_l} \right\} R(j\omega)X)\|_\infty.$$

Since $\sum_{r:l\in r} \hat{x}_r = \hat{y}_l$, it follows that $\|diag\{\frac{1}{\hat{y}_l}\}R(j\omega)X)\|_\infty = 1$. It is also easy to see that

$$\|\mathcal{K}W diag\{\frac{1}{\hat{q}_r}\}R^T(-j\omega)F_p diag\{\hat{y}_l\}\|_\infty = \kappa_r w_r T_r \frac{\sum_{l:l\in r} f_l'(\hat{y}_l)\hat{y}_l}{\hat{q}_r}.$$

Now, we are ready to prove the following theorem [94].

Theorem 6.2. *If*

$$\kappa_r w_r T_r \frac{\sum_{l:l\in r} \hat{p}_l' \hat{y}_l}{\hat{q}_r} \leq \frac{\pi}{2},$$

then the linearized dynamics of the network with the congestion controllers given by (6.7) is stable.

Proof. Define $\tilde{Q}(s) = \sqrt{M_1}R^T(-j\omega)F_pR(j\omega))\sqrt{M_1}$. Note that

$$\sigma\left(\tilde{Q}(j\omega)\tilde{\mathcal{D}}(j\omega)\right) = \sigma\left(\hat{R}(j\omega)\tilde{\mathcal{D}}(j\omega)\hat{R}^T(-j\omega)\right),$$

where

$$\hat{R}(j\omega) = \sqrt{F_p}R^T(j\omega)\sqrt{M_1}. \qquad (6.12)$$

From the discussion prior to the theorem,

$$\begin{aligned}\lambda_{max}\left(\hat{R}(j\omega)\hat{R}^T(-j\omega)\right) &= \lambda_{max}\left(M_1R^T(-j\omega)F_pR(j\omega)\right)\\ &\leq \max_r \kappa_r T_r w_r\\ &\leq \frac{\pi}{2},\end{aligned}$$

where the last line follows from the conditions of the theorem. Let λ be an eigenvalue of $\hat{R}(j\omega)\tilde{\mathcal{D}}(j\omega)R^T(-j\omega)$ and $\mathbf{v}$ be the corresponding eigenvector such that $\|\mathbf{v}\| = 1$. As in the proof of Theorem 6.1, we have

$$\lambda(\tilde{Q}\tilde{\mathcal{D}}) = \mathbf{v}^*\hat{R}(j\omega)\tilde{\mathcal{D}}(j\omega)R^T(-j\omega)\mathbf{v}.$$

Let $\mathbf{u} := R^T(-j\omega)\mathbf{v}$. Since

$$\|R^T(-j\omega)\mathbf{v}\| \leq \lambda_{max}\left(\hat{R}(j\omega)\hat{R}^T(-j\omega)\right) \leq \frac{\pi}{2},$$

by Lemma 5.1,

$$\mathbf{u}^*\tilde{\mathcal{D}}\mathbf{u} = \sum_r |u_r|^2 \frac{e^{-j\omega T_r}}{j\omega T_r + \alpha_r}$$

does not encircle -1. Thus, the result follows from the multivariable Nyquist criterion. □

Again, the choice of $\{\kappa_r\}$ is decentralized; each source r only needs information about its path to choose κ_r. Theorem 6.2 also provides insight into the interplay between the design of the marking functions $\{f_l\}$ at the links and the choice of the controller gains $\{\kappa_r\}$ at the sources. Suppose we assume that

$$\hat{p}'_l\hat{y}_l \leq B\hat{p}_l,$$

then κ_r should be chosen to satisfy $\kappa_r w_r T_r \leq \pi/2B$.

In our definition of high-throughput TCP, we have assumed that user r has the utility function $w_r \log x_r$. The result in Theorem 6.2 can be generalized to the case where user r's utility function is some concave function $U_r(x_r)$. The congestion controller is then given by

$$\dot{x}_r = \kappa_r x_r(t - T_r)\left(1 - \frac{q_r(t)}{U'_r(x_r)}\right).$$

In this case, the stability condition is given by

$$\kappa_r T_r \frac{\sum_{l:l\in r} \hat{p}'_l \hat{y}_l}{\hat{q}_r} \leq \frac{\pi}{2}.$$

In the next two subsections, we show that the condition in Theorem 6.2 is also sufficient for the stability of high-throughput TCP when exponential smoothing is used to estimate the arrival rate at each link, and when the congestion feedback from each link is provided by probabilistically marking each packet.

6.1.3 Exponentially smoothed rate feedback

We continue to consider high-throughput TCP. However, the price at each link is no longer a function of the instantaneous rate at the link, but is rather a function of an estimated rate. As before,

$$\mathbf{x}(s) = M_1(s)M_2(s)\mathbf{q}(s).$$

However, since the link price p_l for each link l is computed as a function of the estimated rate z_l, we have

$$\mathbf{p}(s) = F_p(s)\mathbf{z}(s).$$

Recall that z_l is computed according to the differential equation

$$\varepsilon_l \dot{z}_l = -z_l + y_l.$$

Thus,

$$\mathbf{z}(s) = diag\{\frac{1}{\varepsilon_l s + 1}\}\mathbf{y}(s) + \varepsilon \mathbf{z}_0,$$

and

$$\mathbf{p}(s) = F_p diag\{\frac{1}{\varepsilon_l s + 1}\}\mathbf{y}(s).$$

The loop transfer function is given by

$$I + \tilde{D}(s)M_1 R^T(-s)diag\{\frac{\hat{p}'_l}{1 + \varepsilon_l s}\}R(s),$$

where M_1 and $\tilde{D}$ were defined in (6.9) and (6.11), respectively. Next, note that

$$\sigma\left(\tilde{D}(s)M_1 R^T(-s)diag\{\frac{\hat{p}'_l}{1 + \varepsilon_l s}\}R(s)\right) = \sigma\left(\tilde{D}(s)\hat{R}(-s)diag\{\frac{1}{1 + \varepsilon_l s}\}\hat{R}(s)\right),$$

where $\hat{R}(s)$ was defined in (6.12). Further,

$$\sigma\left(\tilde{D}(s)\hat{R}^T(-s)diag\{\frac{1}{1 + \varepsilon_l s}\}\hat{R}(s)\right) = \sigma\left(diag\{\frac{1}{1 + \varepsilon_l s}\}\hat{R}(s)\tilde{D}(s)\hat{R}^T(-s)\right).$$

Let λ be an eigenvalue of

$$\tilde{D}(j\omega)M_1R^T(-j\omega)diag\{\frac{\hat{p}_l'}{1+j\omega\varepsilon_l}\}R(j\omega),$$

and $\mathbf{v}$ be the corresponding normalized eigenvector. From the above discussion, λ is also an eigenvalue of

$$diag\{\frac{1}{1+\varepsilon_l s}\}\hat{R}(s)\tilde{D}(s)\hat{R}^T(-s).$$

Thus,

$$\lambda\mathbf{v} = diag\{\frac{1}{1+\varepsilon_l s}\}\hat{R}(s)\tilde{D}(s)\hat{R}^T(-s)\mathbf{v},$$

where $\mathbf{v}$ is the normalized eigenvector corresponding to λ. It is easy to see that

$$\lambda = \frac{\mathbf{v}^*\hat{R}(j\omega)\tilde{D}(j\omega)\hat{R}^T(-j\omega)\mathbf{v}}{\mathbf{v}^*diag\{1+j\omega\varepsilon_l\}\mathbf{v}}.$$

Now, suppose that

$$\kappa_r w_r T_r \frac{\sum_{l:l\in r}\hat{p}_l'\hat{y}_l}{\hat{q}_r} \leq 1,$$

a condition that is slightly more restrictive than the assumption in Theorem 6.2. Letting $\mathbf{u} := \hat{R}^T(-j\omega)\mathbf{v}$, we know that $\|\mathbf{u}\| \leq 1$ since $\|\hat{R}^T(-j\omega)\| \leq 1$ due to the above condition on $\{\kappa_r\}$. Thus,

$$\lambda = \frac{\sum_r |u_r|^2 \frac{e^{-j\omega T_r + \alpha_r}}{j\omega T_r}}{1+j\omega\sum_l \varepsilon_l|v_l|^2}.$$

The following two lemmas show that the above expression is greater than -1. The first of the two lemmas shows that the numerator is greater than -1 and the next lemma proves that $\lambda > -1$.

Lemma 6.3. *If $\alpha > 0$, then*

$$\Re\left(\frac{e^{-jx}}{jx+\alpha}\right) > -1$$

for all $x \in (-\infty, \infty)$.

Proof.

$$\frac{e^{-jx}}{jx+\alpha} = \frac{\cos x - j\sin x}{\alpha + jx}$$

$$= \frac{(\cos x - j\sin x)(\alpha - jx)}{\sqrt{\alpha^2 + x^2}}$$

$$= \frac{\alpha\cos x - x\sin x - j(\alpha\sin x + x\cos x)}{\sqrt{\alpha^2 + x^2}}.$$

Thus,

$$\Re\left(\frac{e^{-jx}}{jx+\alpha}\right) = \cos\theta\cos x - \sin\theta\sin x$$
$$= \cos(\theta + x) > -1,$$

where θ is the phase of $\alpha + jx$. □

Lemma 6.4. *Suppose that* $\alpha \geq 1$. *Then*

$$\Re\left(\frac{\alpha + j\beta}{1 + j\delta}\right) > -1,$$

whenever

$$\Im\left(\frac{\alpha + j\beta}{1 + j\delta}\right) = 0.$$

Proof. Note that

$$\frac{\alpha + j\beta}{1 + j\delta} = \frac{(\alpha + \beta\delta) + j(\beta - \alpha\delta)}{1 + \delta^2}.$$

Thus,

$$\Im\left(\frac{\alpha + j\beta}{1 + j\delta}\right) = 0$$

implies that $\beta = \alpha\delta$, which further implies that

$$\Re\left(\frac{\alpha + j\beta}{1 + j\delta}\right) = \alpha > -1.$$

□

The above discussion is summarized in the following theorem.

Theorem 6.5. *The linearized dynamics of the network of high-throughput TCP congestion with the link congestion feedback based on an exponentially smoothed estimate of the arrival rate at the link is stable if*

$$\kappa_r w_r T_r \frac{\sum_{l:l\in r} \hat{p}_l' \hat{y}_l}{\hat{q}_r} \leq 1.$$

□

6.1.4 High-throughput TCP with probabilistic marking

Suppose that the price information is conveyed in a probabilistic manner. In other words, suppose that p_l is the probability that a packet is marked at link l, and q_r is the probability that a packet is marked on a route r. Then,

$$q_r(t) = 1 - \prod_{l:l\in r}(1 - p_l(t - d_2(l,r))).$$

Thus, q_r is no longer the sum of the link prices on route r. Linearizing the relationship between q_r and the p_l's yields

$$\delta q_r(t) = \sum_{l:l \in r} \frac{1 - \hat{q}_r}{1 - \hat{p}_l} \delta p_l(t - d_2(l, r)),$$

where

$$\hat{q}_r = 1 - \prod_{l:l \in r} (1 - \hat{p}_l).$$

Thus,

$$\mathbf{q}(s) = diag\{1 - \hat{q}_r\} R^T(-s) diag\{\frac{1}{1 - \hat{p}_l}\} \mathbf{p}(s).$$

The loop transfer function then becomes

$$I + \tilde{D}(s) M_1 diag\{1 - \hat{q}_r\} R^T(-s) diag\{\frac{1}{1 - \hat{p}_l}\} F_p R(s).$$

Note that

$$\sigma\left(M_1 diag\{1 - \hat{q}_r\} R^T(-s) diag\{\frac{1}{1 - \hat{p}_l}\} F_p R(s)\right)$$

$$= \sigma\left(\mathcal{K} W diag\{\frac{1 - \hat{q}_r}{\hat{q}_r}\} R^T(-j\omega) diag\{\frac{\hat{p}'_l}{1 - \hat{p}_l}\} R(j\omega) X\right)$$

$$= \|\mathcal{K} W diag\{\frac{1 - \hat{q}_r}{\hat{q}_r}\} R^T(-j\omega) diag\{\frac{\hat{p}'_l \hat{y}_l}{1 - \hat{p}_l}\}\|_\infty \|diag\{\frac{1}{y_l}\} R(j\omega) X\|_\infty.$$

As before,

$$\|diag\{\frac{1}{y_l}\} R(j\omega) X\|_\infty \leq 1.$$

Further,

$$\|\mathcal{K} W diag\{\frac{1 - \hat{q}_r}{\hat{q}_r}\} R^T(-j\omega) diag\{\frac{\hat{p}'_l \hat{y}_l}{1 - \hat{p}_l}\}\|_\infty$$

$$= \max_r \kappa_r w_r T_r \frac{(1 - \hat{q}_r)}{\hat{q}_r} \sum_{l:l \in r} \frac{\hat{p}'_l \hat{y}_l}{1 - \hat{p}_l}.$$

Now, by mimicking the proof of Theorem 6.2, we have the following result.

Theorem 6.6. *The network of congestion controllers with high-throughput TCP at the sources, probabilistic feedback from the routers, and the marking probability at each link being a function of an exponentially smoothed estimate of the rate is stable if*

$$\kappa_r w_r T_r \frac{(1 - \hat{q}_r)}{\hat{q}_r} \sum_{l:l \in r} \frac{\hat{p}'_l \hat{y}_l}{1 - \hat{p}_l} < \frac{\pi}{2}.$$

$\square$

If $\hat{p}_l' \hat{y}_l \leq B\hat{p}_l$, then $\kappa_r w_r T_r < \pi/2B$ is sufficient to ensure local stability. To see this, we note that

$$\kappa_r w_r T_r \frac{(1 - \hat{q}_r)}{\hat{q}_r} \sum_l \frac{\hat{p}_l' \hat{y}_l}{1 - \hat{p}_l} < \kappa_r w_r T_r B \frac{(1 - \hat{q}_r)}{\hat{q}_r} \sum_{l:l \in r} \frac{\hat{p}_l}{1 - \hat{p}_l}$$

$$= \kappa_r w_r T_r B \frac{1}{\hat{q}_r} \sum_{l:l \in r} \hat{p}_l \prod_{m:m \in r, m \neq l} (1 - \hat{p}_m).$$

Note that

$$\sum_{l:l \in r} \hat{p}_l \prod_{m:m \in r, m \neq l} (1 - \hat{p}_m)$$

is the probability that a packet is marked exactly in one link on route r, whereas q_r is the probability that a packet is marked in any one of the links on route r. Thus,

$$\frac{1}{\hat{q}_r} \sum_{l:l \in r} \hat{p}_l \prod_{m:m \in r, m \neq l} (1 - \hat{p}_m) \leq 1,$$

giving us the desired result.

6.1.5 A general nonlinear increase/decrease algorithm

Recall that, in the window flow-control implementation of high-speed TCP, the window is incremented by a constant amount for each received acknowledgment. If the RTT for a source-destination pair is small, then this results in a very rapid window increase, thus possibly leading to large oscillations around the equilibrium. On the other hand, in current versions of TCP, the window is incremented by the inverse of the current window size which is not suitable for high-throughput operation over large RTTs and was thus the motivation for introducing high-throughput TCP. Thus, it may be important to strike a balance between these two modes of operation for large and small RTTs. We will discuss this further in a later chapter on stochastic models. For now, we will consider the following the congestion controller introduced in [94] which provides many options for the increase and decrease portions of the congestion avoidance phase of TCP:

$$\dot{x}_r = \kappa_r x_r(t - T_r) \left((1 - q_r) \frac{a_r}{x_r^m} - q_r b_r x_r^n \right). \tag{6.13}$$

This is a generalization of the algorithms proposed in [80, 75, 7]. As compared to other algorithms that we have studied in this chapter, we have introduced a $(1 - q_r)$ term in the increase portion of the algorithm. If we interpret q_r to be the marking probability, then $(1 - q_r)$ is the probability of receiving no mark. Thus, in (6.13), the window size is incremented only when no mark is received.

It is instructive to express (6.13) in terms of window sizes by making the substitution $W_r = T_r x_r$, where W_r is the window size of user r. Upon making this change of variables, the congestion controller for source r becomes

$$\dot{W}_r = \kappa_r \left(\frac{a_r T_r^m}{W_r^m} (1 - q_r) x_r(t - T_r) - \frac{b_r W_r^n}{T_r^n} q_r x_r(t - T_r) \right). \tag{6.14}$$

Interpreting $q_r x_r(t - T_r)$ as the rate at which marks are received and $(1 - q_r) x_r(t - T_r)$ as the rate at which acknowledgments are received for unmarked packets, it should be clear that the above algorithm increases the window size by $a_r d_r^m / W_r^m$ for each unmarked ack and decreases it by $b_r W_r^n / d_r^n$ for each mark. By choosing a_r and b_r, both Jacobson's algorithm in current versions of TCP and the high-throughput TCP congestion controller can be considered as special cases of (6.14). Now, we proceed to the stability analysis of (6.13).

Linearizing (6.13), we obtain

$$\delta \dot{x}_r = \kappa_r \hat{x}_r \left(-\frac{a_r}{\hat{x}_r^m} \delta q_r - (1 - \hat{q}_r) \frac{m a_r}{\hat{x}_r^{m+1}} \delta x_r - b_r \hat{q}_r n \hat{x}_r^{n-1} \delta x_r - b_r \hat{x}_r^n q_r \right), \tag{6.15}$$

where, from (6.13), the equilibrium values of x_r and q_r are related by the equation

$$\frac{(1 - \hat{q}_r) q_r}{\hat{x}_r^{m+1}} = \hat{q}_r b_r \hat{x}_r^{n-1}. \tag{6.16}$$

Taking the Laplace transform of (6.15) and using (6.16), we get

$$x_r(s) = -\frac{\kappa_r T_r \hat{x}_r}{\hat{q}_r} \frac{a_r}{\hat{x}_r^m} \frac{1}{sT_r + \alpha_r} q_r(s), \tag{6.17}$$

where α_r is now given by

$$\alpha_r = \kappa_r T_r b_r \hat{x}_r^n (m + n).$$

Comparing (6.17) to (6.8), we notice that the role of w_r is now played by $\frac{a_r}{\hat{x}_r^m}$. Further, $\alpha_r > 0$ if $m + n > 0$. Thus, we have the following analog of Theorems 6.2 and 6.6.

Theorem 6.7. *Consider the congestion controller given by (6.15) with $m + n > 0$.*

- *If the feedback is price-based, i.e.,*

$$q_r(t) = \sum_{l:l \in r} p_l(t - d_2(l, r)),$$

then

$$\kappa_r T_r a_r \hat{x}_r^m \frac{\sum_{l:l \in r} f_l'(\hat{y}_l) \hat{y}_l}{\hat{q}_r} < \frac{\pi}{2}$$

is a sufficient condition for the linearized version of (6.15) to be asymptotically stable.

- If the feedback is probabilistic, i.e.,

$$1 - q_r(t) = \prod_{l:l \in r} [1 - p_l(t - d_2(l,r))],$$

then

$$\kappa_r T_r \frac{a_r}{\hat{x}_r^m} \frac{1 - \hat{q}_r}{\hat{q}_r} \sum_{l:l \in r} \frac{\hat{p}_l' \hat{y}_l}{1 - \hat{p}_l} < \frac{\pi}{2}$$

is a sufficient condition for the linearized version of (6.15) to be asymptotically stable.

Further, if $f_l'(\hat{y}_l)\hat{y}_l \leq B f_l(\hat{y}_l) \; \forall l$, then

$$\kappa_r T_r \frac{a_r}{\hat{x}_r^m} < \frac{\pi}{2B}$$

is a sufficient condition for the linearized version of (6.15) to be asymptotically stable, independent of whether the feedback is price-based or probabilistic. □

6.1.6 High-throughput TCP and AVQ

Consider the high-throughput TCP congestion controller

$$\dot{x}_r(t) = k_r x_r(t - T_r) \left(1 - \frac{q_r(t)}{U_r'(x_r(t))} \right) \tag{6.18}$$

along with the link price equation

$$p_l = f_l(y_l, \tilde{c}_l) \tag{6.19}$$

where we have explicitly stated the dependence of the link price on $\tilde{c}_l$, the virtual capacity of link l. Recall that, in the AVQ algorithm, c_l is adapted according to

$$\dot{\tilde{c}}_l = \alpha_l(\gamma_l c_l - y_l). \tag{6.20}$$

Let $p_y^l = \frac{\partial p_l}{\partial y_l} > 0$ and $p_c^l = \frac{\partial p_l}{\partial \tilde{c}_l} < 0$. Linearizing (6.18) and (6.20) around the equilibrium point and taking the Laplace transform of the resulting equations, we get

$$s\tilde{c}_l(s) = -\alpha_l y_l(s),$$

$$p_l(s) = p_y^l y_l(s) + p_c^l \tilde{c}_l(s) = (p_y^l - \frac{\alpha_l}{s} p_c^l) y_l(s)$$

$$= p_y^l (1 + \frac{\beta_l}{s}) y_l(s),$$

where $\beta_l = -\alpha_l p_c^l / p_y^l$

As in the case of other controllers considered earlier in this chapter, it is easy to show that the loop transfer function is given by

$$L(s) = \text{diag}\left(k_r T_r \frac{\hat{x}_r}{\hat{q}_r} \frac{e^{-sT_r}}{sT_r + \alpha_r}\right) R^T(-s)\text{diag}(p_y^l(1 + \frac{\beta_l}{s}))R(s).$$

By the multivariable Nyquist criterion, the closed loop system is asymptotically stable if the eigenvalues of $L(j\omega)$ do not encircle -1. Now the eigenvalues of $L(j\omega)$ are identical to those of

$$\hat{G}(j\omega) = \text{diag}\left(\frac{e^{-j\omega T_r}}{j\omega T_r + \alpha_r}\right) \hat{R}^T(-j\omega)\text{diag}\left(1 + \frac{\beta_l}{s}\right) \hat{R}(j\omega)).$$

where

$$\hat{R}(j\omega) = \text{diag}\left(\sqrt{p_y^l}\right) R(j\omega)\text{diag}\left(\sqrt{k_r T_r \frac{\hat{x}_r}{\hat{q}_r}}\right).$$

Let $T_m = \max_{r \in R} T_r$, and let $\beta_l = \frac{n_l}{T_m}$, then we have the following result [67].

Theorem 6.8. *A network of high-throughput TCP congestion controllers at the sources and AVQ at the links is stable around its equilibrium if the following conditions are satisfied:*

1. $k_r \le \frac{1}{T_r} \min_{l \in r} \frac{\hat{p}_l}{p_y^l c_l}$, *and*
2. $n_l \le 0.49$ *for all* $l \in L$.

Proof. Under condition (1), it is easy to show that the norm of $\hat{R}(s)$ is smaller than unity, so for any eigenvalue λ of $G(s)$, we can find two vectors v, u, the norm of v is 1 and the norm of u is smaller than 1, such that:

$$\lambda = \frac{num}{den} = \frac{u^* \text{diag}(\frac{e^{-j\omega T_r}}{j\omega T_r + \alpha_r})u}{v^* \text{diag}(\frac{j\omega}{j\omega + \beta_l})v}.$$

Let num denote the numerator and den denote the denominator of the right-hand side of the above expression. We will assume that $\omega \ge 0$. The case $\omega < 0$ can be handled in a similar manner.

We now consider two cases.

(i) *Suppose that* $\omega > \frac{1}{T_m}$.

In this case,

$$1 + \frac{\beta_l}{s} = 1 - \frac{n_l}{\omega T_m}j.$$

So

$$1 < \left|1 + \frac{\beta_l}{j\omega}\right| < \sqrt{1 + n_l^2},$$

and

$$0 > Phase\left(1 + \frac{\beta_l}{j\omega}\right) > -\arctan n_l.$$

Alternatively, we have

$$\frac{1}{\sqrt{1+n_l^2}} < \left| \frac{j\omega}{j\omega + \beta_l} \right| < 1$$

and

$$0 < Phase\left(\frac{j\omega}{j\omega + \beta_l} \right) < \arctan n_l.$$

Let $n_m = max_{l \in L} n_l$, then

$$\frac{1}{\sqrt{1+n_m^2}} < \left| \frac{j\omega}{j\omega + \beta_l} \right| < 1,$$

and

$$0 < Phase\left(\frac{j\omega}{j\omega + \beta_l} \right) < \arctan n_m.$$

The above inequalities are true for any $l \in L$, so we have

$$\frac{1}{\sqrt{1+n_m^2}} < |den| < 1 \ \ \text{and} \ \ 0 < Phase(den) < \arctan n_m.$$

Let $Co(S)$ denote the convex hull of a set S, i.e.,

$$Co(S) = \left\{ \sum_i \alpha_i x_i : x_i \in S, \alpha_i \geq 0 \ \ \forall i \ \text{and} \ \sum_i \alpha_i = 1 \right\}.$$

Thus, λ lies in

$$\sqrt{1+n_m^2} e^{-j \arctan n_m} Co\left(0 \cup \{ \frac{e^{-jx}}{jx+\alpha} \} \right),$$

where $\alpha = \min_r \alpha_r$. The imaginary part of

$$e^{-j \arctan n_m} \frac{e^{-jx}}{jx+\alpha}$$

is equal to zero when

$$x + Phase(jx+\alpha) + \arctan n_m = 0, \pm\pi, \pm 2\pi, \ldots.$$

For these values of the imaginary part, the real part is closest to -1 when

$$x = \pi - Phase(jx+\alpha) - \arctan n_m.$$

The magnitude of the real part at these values is upper-bounded as

$$|Re(G(j\omega)| \leq \frac{\sqrt{1+n_m^2}}{\sqrt{(\pi - \theta - \arctan n_m)^2 + \alpha^2}},$$

where
$$\theta = Phase(jx + \alpha).$$

Noting that $\theta < \pi/2$, it is easy to see that a sufficient condition for $Re(G(j\omega)) > -1$ when $Im(G(j\omega)) = 0$ is given by

$$\sqrt{1 + n_m^2} + \arctan n_m < \pi/2.$$

It can be numerically verified that this condition holds when $n_m \leq 0.49$. This is illustrated in Figures 6.1 and 6.2.

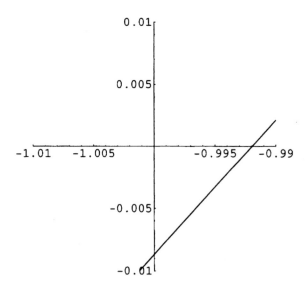

Fig. 6.1. Plot of $\sqrt{1 + n_m^2}\, e^{-j \arctan n_m}\, \frac{e^{-j\omega}}{j\omega + 0.005}$ for $n_m = 0.49$ as a function of ω, around the point -1. It can be seen that the plot crosses the real axis to the right of -1.

(ii) Suppose that $\omega \leq \frac{1}{T_m}$.

In this case, since

$$Phase(\frac{e^{-j\omega T_r}}{j\omega T_r + \alpha_r}) > -\omega T_r - \arctan \frac{\omega T_r}{\alpha_r},$$

we have

$$Phase(num) > -\omega T_m - \arctan \frac{\omega T_m}{min_{r \in R} \alpha_r}.$$

Since

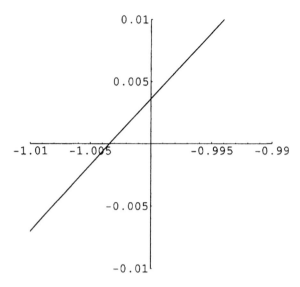

Fig. 6.2. Plot of $\sqrt{1 + n_m^2}\, e^{-j\arctan n_m}\, \frac{e^{-j\omega}}{j\omega + 0.005}$ for $n_m = 0.5$ as a function of ω, around the point -1. It can be seen that the plot crosses the real axis to the left of -1, thus admitting the possibility of encircling -1.

$$Phase\left(\frac{j\omega}{j\omega + \beta_l}\right) = \frac{\pi}{2} - \arctan\frac{\omega T_m}{n_l},$$

we have

$$Phase(den) < \frac{\pi}{2} - \arctan\frac{\omega T_m}{n_m}.$$

So,

$$Phase(\lambda) > -\omega T_m - \arctan\frac{\omega T_m}{min_{r \in R}\alpha_r} - \frac{\pi}{2} + \arctan\frac{\omega T_m}{n_m}.$$

If $\omega T_m < \arctan\frac{\omega T_m}{n_m}$, since $\arctan\frac{\omega T_m}{min_{r \in R}\alpha_r} < \pi/2$, we have $Phase(\lambda) > -\pi$, and λ will always be in the lower half of the complex plane, i.e., it cannot encircle -1. Thus, we require $\omega T_m < \arctan\frac{\omega T_m}{n_m}$, or equivalently,

$$\frac{\tan \omega T_m}{\omega T_m} < \frac{1}{n_m}$$

when $\omega T_m < 1$. This puts an upper-bound on n_m given by

$$n_m \leq \frac{1}{\tan 1} \approx 0.64.$$

So we know that for Case (i), we need $n_m \leq 0.49$, and for Case (ii), we need $n_m \leq 0.64$ to ensure that the eigenlocus of $G(j\omega)$ does not encircle -1. Thus, $n_m < 0.49$, is a sufficient condition for stability.   □

We note that a similar stability condition was derived first in [60, 59, 63] for the case of proportionally-fair congestion controllers with AVQ-based congestion feedback.

6.2 Dual algorithm

Recall that the source control in the dual algorithm is a static law given by

$$U'(x_r) = q_r. \tag{6.21}$$

Following [83], linearizing around the equilibrium point gives

$$U''(\hat{x}_r)\delta x_r(t) = \delta q_r(t).$$

In the Laplace transform domain, we get

$$\mathbf{x}(s) - \mathbf{x}_0 = diag\{\frac{1}{U_r''(\hat{x}_r)}\}\mathbf{q}(s) = diag\{\frac{-1}{|U_r''(\hat{x}_r)|}\}D(s)R^T(-s)\mathbf{p}(s).$$

Suppose that we use the following dynamics to generate the link price:

$$\dot{p} = \frac{1}{c_l}[y_l - c_l]_0^{y_l}. \tag{6.22}$$

Note that p_l can be interpreted as the delay at link l. Linearizing and taking the Laplace transform gives

$$\mathbf{p}(s) - \mathbf{p}_0 = diag\{\frac{1}{s}\}diag\{\frac{1}{c_l}\}\mathbf{y}(s).$$

Finally, recall that $\mathbf{p}(s) = R(s)\mathbf{x}(s)$. Thus, we get the network transfer function

$$\left[I + D(s)diag\{\frac{T_r}{|U_r''(\hat{x}_r)|}|\}R^T(-s)diag\{\frac{1}{c_l}\}R(s)\right]\mathbf{x}(s)$$

$$= \mathbf{x}(0) + diag\{\frac{1}{|U_r''(\hat{x}_r)|}\}D(s)R^T(-s)\mathbf{p}_0.$$

Consider the matrix defined by

$$Q(s) = diag\{\frac{T_r}{|U_r''(\hat{x}_r)|}\}R^T(-s)diag\{\frac{1}{c_l}\}R(s).$$

Note that

$$\sigma(Q(s)) = \sigma\left(diag\{\sqrt{\frac{T_r}{|U_r''(\hat{x}_r)|}}\}R^T(-s)diag\{\frac{1}{c_l}\}R(s)diag\{\sqrt{\frac{T_r}{|U_r''(\hat{x}_r)|}}\}\right).$$

The matrix on the right-hand side of the above equation is positive semi-definite. Thus, the eigenvalues of $Q(s)$ are non-negative. Further,

$$\sigma(Q(s)) = \sigma(X^{-1}diag\{\frac{T_r}{|U_r''(\hat{x}_r)|}\}R^T(-s)diag\{\frac{1}{c_l}\}R(s)X).$$

Thus,

$$\lambda_{max}(Q(s)) \leq \|X^{-1}diag\{\frac{T_r}{|U_r''(\hat{x}_r)|}\}R^T(-s)\|_\infty \|diag\{\frac{1}{c_l}\}R(s)X\|_\infty.$$

As before, if we can show that $\lambda_{max}(Q(s)) < \pi/2$, then by the multivariable Nyquist criterion, the network will be locally stable. Since $\sum_{l:l\in r} x_r \leq c_l$, it follows that

$$\|diag\{\frac{1}{c_l}\}R(s)X\|_\infty \leq 1.$$

Choosing

$$\frac{1}{|U_r''(\hat{x}_r)|} = \frac{\kappa_r x_r}{T_r L_r},$$

where L_r is the number of links in route r, and $k_r \leq \pi/2$ is a constant, it is easy to see that

$$\|X^{-1}diag\{\frac{T_r}{|U_r''(\hat{x}_r)|}\}R^T(-s)\|_\infty \leq \frac{\pi}{2},$$

proving the following result.

Theorem 6.9. *A network of dual source controllers given by (6.21) along with the link price dynamics (6.22) is stable around its equilibrium if the following conditions are satisfied:*

- $\frac{1}{|U_r''(\hat{x}_r)|} = \frac{\kappa_r x_r}{T_r L_r}$, *and*
- $\kappa_r \leq \pi/2$, $\forall r$.

$\square$

In order to establish the above result, we required the source law to be

$$U''(x_r) = -\frac{T_r L_r}{\kappa_r x_r}.$$

A particular utility function which satisfies this second derivative constraint is

$$U_r(x_r) = \frac{T_r L_r}{\kappa_r} x_r \left(1 - \log \frac{x_r}{x_{r,max}}\right), \tag{6.23}$$

where $x_{r,max}$ is an upper-bound on x_r, i.e., it is assumed that $\hat{x}_r \leq x_{r,max}$. Thus, source r's congestion-control algorithm becomes

$$x_r = x_{r,max} e^{-\frac{\kappa_r q_r}{T_r L_r}}. \tag{6.24}$$

Note that the utility function (6.23) is biased by the presence of the RTT in the utility function. Therefore, unlike the penalty function-based method, to ensure the stability of the dual algorithm in the presence of feedback delays, one has to sacrifice the ability to allocate resources in an arbitrary manner. There is a way to fix this problem as we will show later.

6.2.1 Fair dual algorithm

The AVQ algorithm was designed for the primal algorithm, which asymptotically satisfies any arbitrary fairness criterion, to achieve full utilization. The dual algorithm in the previous section achieves full utilization, but is not able to realize arbitrary fairness criteria. Specifically, the utility function is restricted to be of the form (6.23). Borrowing the key idea from the AVQ algorithm which slowly adapts the links' AQM parameters to achieve full utilization, it is natural to attempt to slowly adapt the sources' control algorithms to achieve fairness in the dual algorithm. We now present the algorithm in [84], where the parameter $x_{r,max}$ in (6.24) is adapted to achieve fairness.

Suppose that $x_{r,max}$ is chosen dynamically according to the following equations:

$$x_{r,max} = x_{m,r}e^{\xi_r}, \tag{6.25}$$

$$T_r\dot{\xi}_r = \beta_r\left(U_r'(x_r) - q_r\right), \tag{6.26}$$

where $x_{m,r}$ is some positive constant. In equilibrium, it is clear that $U_r'(x_r)$ will be equal to q_r, thus achieving fair resource allocation. Thus, our goal is to choose $\{\beta_r\}$ to ensure that the system converges to the equilibrium point.

Linearizing (6.25)-(6.26) yields

$$\delta x_r = \hat{x}_r\delta\xi_r - \frac{\kappa_r\hat{x}_r}{T_rL_r}\delta q_r,$$

$$T_r\delta\dot{\xi}_r = \beta_r\left(U_r''(\hat{x}_r)\delta x_r - \delta q_r\right).$$

Taking Laplace transforms and setting initial conditions equal to zero , we get

$$x_r(s) = -\frac{\kappa_r\hat{x}_r}{T_rL_r}q_r(s) + \hat{x}_r\xi_r(s),$$

$$\xi_r(s) = \frac{\beta_rU_r''(\hat{x}_r)}{sT_r}x_r(s) - \frac{\beta_r}{sT_r}q_r(s).$$

Thus,

$$x_r(s) = -\alpha_1\frac{s + \alpha_2}{s + \alpha_3},$$

where

$$\alpha_1 := \frac{\kappa_r\hat{x}_r}{L_r}, \qquad \alpha_2 := \frac{\beta_rL_r}{\kappa_r}, \qquad \alpha_3 = \frac{\beta_r\hat{x}_r|U_r''(\hat{x}_r)|}{T_r}.$$

Before we proceed further, we state the following lemma.

Lemma 6.10. *Let T_{max} be the maximum RTT of all the sources in the network, and choose β_r such that*

$$\frac{\beta_r L_r}{\kappa_r} \leq \frac{T_{max}}{\tan(1)} \approx 0.64 T_{max}.$$

The function $\frac{e^{-j\omega T_r}}{j\omega T_r} \frac{j\omega + \alpha_2}{j\omega + \alpha_3}$ does not encircle the point -1 as ω is varied from $(-\infty, +\infty)$.

Proof. The proof is very similar to the proof of Theorem 6.8 and is therefore not provided here.   □

Now, proceeding as before, we can show that the system is stable if $\kappa_r < \pi/2$ and $\frac{\beta_r L_r}{\kappa_r} < 0.64 T_{max}$. Thus, as in AVQ, where the adaptation parameters at the links had to be inversely proportional to the longest RTT in the network, here the adaptation parameters $\{\beta_r\}$ have to be inversely proportional to the longest RTT in the network. The stability result for the fair dual algorithm is summarized in the following theorem.

Theorem 6.11. *A network of dual source controllers given by (6.25)-(6.26) along with the link price dynamics (6.22) is stable around its equilibrium if the following conditions are satisfied:*

- $\kappa_r \leq \pi/2$, *and*
- $\frac{\beta_r L_r}{\kappa_r} \leq 0.64 T_{max}$, $\forall r$.

□

6.3 Primal-dual algorithm

Recall the primal-dual algorithm, with the source controller

$$\dot{x}_r(t) = \left[\kappa_r(x_r(t), x_r(t - T_r))(1 - \frac{q_r(t)}{U'_r(x_r(t))})\right]^+_{x_r}, \qquad (6.27)$$

and the link price updated according to

$$\dot{p}_l(t) = \left[h_l(p_l(t))(y_l(t) - c_l)\right]^+_{p_l}, \qquad (6.28)$$

where, by slightly abusing notation, we allow κ_r to be a function of both $x_r(t)$ and $x_r(t - T_r)$.

We first linearize (6.27) around its equilibrium point, $\hat{q}_r = U'_r(\hat{x}_r)$. This gives

$$\delta\dot{x}_r(t) = -\hat{\kappa}_r \left(\frac{1}{U'_r(\hat{x}_r)}\delta q_r(t) + \frac{-\hat{q}_r}{(U'_r(\hat{x}_r))^2}U''_r(\hat{x}_r)\delta x_r(t)\right)$$

$$= -\hat{\kappa}_r \left(\frac{1}{\hat{q}_r}\delta q_r(t) + \frac{-1}{\hat{q}_r}U''_r(\hat{x}_r)\delta x_r(t)\right),$$

where $\hat{\kappa}_r = \kappa_r(\hat{x}_r, \hat{x}_r)$.

In the Laplace domain, we have

$$sx_r(x) = -\frac{\hat{\kappa}_r}{\hat{q}_r}q_r(s) + \frac{\hat{\kappa}_r U_r''(\hat{x}_r)}{\hat{q}_r}x_r(s).$$

So, we have

$$x_r(s) = -\frac{\hat{\kappa}_r}{\hat{q}_r}\frac{1}{s - \hat{\kappa}_r U_r''(\hat{x}_r)/\hat{q}_r}q_r(s) = -\frac{1}{u_r''(\hat{x}_r)}\frac{\theta_r}{sT_r + \theta_r}q_r(s), \qquad (6.29)$$

where

$$\theta_r = \frac{\hat{\kappa}_r T_r(-U_r''(\hat{x}_r))}{\hat{q}_r}. \qquad (6.30)$$

Note that $\theta_r > 0$, since $U_r''(x_r) < 0$ due to the concavity of $U_r(x_r)$.

For the link dynamics, linearization yields:

$$\dot{p}_l = h_l(\hat{p}_l)y_l.$$

In the Laplace domain, we have:

$$p_l(s) = \frac{h_l(\hat{p}_l)}{s}y_l(s). \qquad (6.31)$$

The loop transfer function of the feedback system (6.29) and (6.31) is given by

$$G(s) = \text{diag}\left(\frac{1}{-U_r''(\hat{x}_r)}\frac{\theta_r e^{-sT_r}}{sT_r + \theta_r}\right)R^T(-s)\text{diag}\left\{\frac{h_l(\hat{p}_l)}{s}\right\}R(s). \qquad (6.32)$$

First notice that we can rewrite $G(s)$ as:

$$G(s) = \text{diag}\left(\frac{T_r}{-U_r''(\hat{x}_r)}\frac{\theta_r e^{-sT_r}}{sT_r(sT_r + \theta_r)}\right)R^T(-s)\text{diag}(h_l(\hat{p}_l))R(s). \qquad (6.33)$$

For the rest of this section, we will assume that $U_r(x_r)$ is of the form $w_r \log x_r$ or $-\frac{w_r}{x_r^{n_r}}$. If $U_r(x_r) = w_r \log x_r$, we call it a logarithmic utility function; and if $U_r(x_r) = -\frac{w_r}{x_r^{n_r}}$, we call it a power utility function of order n_r.

Theorem 6.12. *The closed loop system described by (6.27) and (6.28) is locally asymptotically stable around the equilibrium point if*

$$\frac{h_l(\hat{p}_l)}{\hat{p}_l} \le \frac{1}{c_l}\min_{r:l\in r}\frac{1}{a_r T_r}, \qquad (6.34)$$

where $a_r := 1$ if $U_r(x_r) = w_r \log x_r$, and $a_r = \frac{1}{1+n_r}$ if $U_r(x_r) = -\frac{w_r}{x_r^{n_r}}$.

Proof. Given the assumption of logarithmic or power utility function and the definition of a_r, we have

$$U_r'' = -\frac{q_r}{a_r x_r}.$$

Let $\beta_l := (h_l(\hat{p}_l)c_l/\hat{p}_l)$. Then from (6.33), we have

$$G(s) = \text{diag}\left(\frac{T_r a_r \hat{x}_r}{\hat{q}_r} \frac{\theta_r e^{-sT_r}}{sT_r(sT_r + \theta_r)}\right) R^T(-s) \text{diag}\left(\frac{\beta_l \hat{p}_l}{c_l}\right) R(s).$$

Since the open-loop system is stable, by the multivariable Nyquist Criterion (see Appendix 6.4), the closed-loop system is asymptotically stable if the eigenloci of $L(j\omega)$ do not encircle -1. Now the eigenvalues of $L(j\omega)$ are identical with those of

$$\hat{G}(j\omega) = \text{diag}\left(\frac{\theta_r e^{-j\omega T_r}}{j\omega T_r(j\omega T_r + \theta_r)}\right) \hat{R}^T(-j\omega)\hat{R}(j\omega),$$

where

$$\hat{R}(j\omega) = \text{diag}\left(\sqrt{\frac{\beta_l \hat{p}_l}{c_l}}\right) R(j\omega) \text{diag}\left(\sqrt{\frac{T_r a_r \hat{x}_r}{\hat{q}_r}}\right).$$

Let $\sigma(Z)$ denote the spectrum of a square matrix Z and $\rho(Z)$ its spectral radius. Given the condition in (6.34), we have:

$$\sigma^2(\hat{R}(j\omega)) = \rho(\hat{R}^T(-j\omega)\hat{R}(j\omega))$$
$$= \rho(\text{diag}\left(\frac{a_r T_r \hat{x}_r}{\hat{q}_r}\right) R^T(-j\omega) \text{diag}\left(\frac{\beta_l \hat{p}_l}{c_l}\right) R(j\omega))$$
$$\leq \|\text{diag}\left(\frac{a_r T_r}{\hat{q}_r}\right) R^T(-j\omega) \text{diag}(\beta_l \hat{p}_l)\| \times \|\text{diag}\left(\frac{1}{\hat{y}_l}\right) R(j\omega) \text{diag}(\hat{x}_r)\|$$
$$\leq 1 \times 1 = 1.$$

The last inequality above uses the facts that

$$\sum_{r:l\in r} \hat{x}_r = \hat{y}_l = c_l, \forall l \in L,$$

and

$$\beta_l < \frac{1}{a_r T_r}, \forall r : l \in r \implies \sum_{l\in r} \hat{p}_l \beta_l \leq \frac{\hat{q}_r}{a_r T_r}, \forall r : l \in r,$$

and so the absolute row sums of these matrices are all less than or equal to 1.

Now, if λ is an eigenvalue of $G(j\omega)$, then

$$\lambda \in \text{Co}(0 \cup \{\frac{\theta_r e^{-j\omega T_r}}{j\omega T_r(j\omega T_r + \theta_r)}\}).$$

Here the reasoning is similar to that in the previous sections. From Lemma 5.6, the plot of

$$\left(\frac{\theta_r e^{-j\omega T_r}}{j\omega T_r (j\omega T_r + \theta_r)} \right)$$

as ω is varied from $-\infty$ to ∞ always crosses the real axis to the right of the point -1. So the eigenloci of $G(j\omega)$ do not encircle -1; thus by the multivariable Nyquist Criterion, the closed-loop system is stable. □

The primal-dual algorithm requires the control gain at the router to be inversely proportional to the maximum round-trip delay of any source using that router. This is a relaxation of the condition for local stability of the AVQ algorithm which requires the gain at the router to be inversely proportional to the maximum round-trip delay in the network. Further, unlike the AVQ algorithm, in a network with only controlled, persistent flows, the router's price computation in the primal-dual algorithm can be interpreted directly in terms of the queue length as we will see later. This interpretation is less clear when there are stochastic disturbances in the network. In [49], Kelly argues that if the queue length hits zero several times within a round-trip time, then the source only sees a stochastically averaged version of the queue length in the deterministic model. However, this result may depend on the modelling assumptions about the system; see [19] for further discussion.

From Theorem 6.12, we see that the local stability depends only on the link price adaptation speed, and not on the source rate adaptation speed. In other words, the stability condition does not depend on $\{\hat{\kappa}_r\}$. If we adapt the source rate infinitely fast (i.e., choosing $\kappa_r(\cdot) \equiv \kappa_r$ and letting $\kappa_r \to \infty$), we get the dual algorithm. Thus, the interesting observation from our analysis is that slow adaptation at the source is not necessary for the dual algorithm to be stable, and achieve full utilization and arbitrary fairness. The dual algorithm can also be stabilized with slow adaptation at the links since our local stability result does not depend on κ_r.

We also note that the price equation in the primal-dual equation is closely related to the RED marking scheme. Consider the link dynamics

$$\dot{p}_l = \beta_l \frac{p_l}{c_l}(y_l - c_l) \quad \text{if} \quad p_l > 0. \tag{6.35}$$

If β is chosen appropriately to satisfy condition (6.34), then the system is stable. The above price equation can be implemented easily by setting the packet marking probability (the link price) p_l to be an exponential function of the queue length b_l:

$$p_l = \begin{cases} 0 & \text{if } 0 \leq \tilde{b}_l < th_{min,l}, \\ p_{min} e^{\frac{\beta_l}{c_l}(\tilde{b}_l - th_{min,l})} & \text{if } th_{min,l} < \tilde{b}_l < th_{max,l}, \\ 1 & \text{if } \tilde{b}_l \geq th_{max,l}, \end{cases} \tag{6.36}$$

where, $\tilde{b}_l = b_l$, and $th_{min,l} < th_{max,l}$ are the two queue-length thresholds between which the exponential marking is selected and p_{min} is the marking threshold when $b_l = th_{min,l}$. Let p_{max} denote the marking probability when

$b_l = th_{max,l}$; then, $th_{min,l}$, $th_{max,l}$, and p_{min} must be such that $p_{max} < 1$. We call the above marking scheme E-RED. Note that E-RED simply uses a different marking profile than RED; specifically, instead of a linear function, we use an exponential function of the queue length. In this section, we have also assumed that the RTT is a constant. To make this applicable to a real network, we have to interpret c_l in (6.35)-(6.36) as a virtual capacity which is smaller than the real capacity. Thus, b_l in (6.36) should be interpreted as the corresponding virtual queue length. This would ensure that the sources get early congestion feedback and therefore, the real queues will not build up significantly.

6.4 Appendix: Multivariable Nyquist criterion

In a single-input, single-output system, the stability of the system is determined by checking if the roots of an equation of the form

$$1 + G(s)H(s) = 0$$

lie in the left-half plane, where $G(s)H(s)$ is some scalar function of the Laplace transform variable s. In the multiple-input, multiple-output case, the corresponding equation is of the form

$$det(I + G(s)H(s)) = 0,$$

where $G(s)$ is now a matrix function that represents the open-loop transfer function, $H(s)$ is the controller transfer function and $I + G(s)H(s)$ is the closed-loop transfer function. The multivariable Nyquist criterion generalizes the Nyquist test given in the previous chapter to the multi-input, multi-output case [21]. Let P denote the number of poles (including multiplicities) of $G(s)H(s)$ in the right-half plane. Let N be the number of counterclockwise encirclements of origin by the spectrum of $G(j\omega)H(j\omega)$ as ω is varied from $-\infty$ to ∞. Then, the closed-loop system is stable if the $N = Z$. A precise statement of this result in given in [21, Theorem L3].

7

Global Stability for a Single Link and a Single Flow

In the previous two chapters, we obtained stability conditions for various congestion control and congestion indication algorithms by linearizing around the equilibrium point. However, such an analysis does not guarantee convergence to equilibrium, starting from an arbitrary initial condition. In this chapter, we consider the global stability problem. However, little is known about the global stability of such controllers in general. Therefore, we confine ourselves to the special case of proportionally-fair controllers for a single link accessing a single link and derive conditions for global stability. While the global stability question is open for general topology networks, the analysis in this chapter and extensive simulations of various controllers in many papers in the congestion-control literature suggest that the conditions obtained from a linear analysis may indeed be sufficient conditions for stability even if the initial conditions lie in a large region of attraction around the equilibrium point.

7.1 Proportionally-fair controller over a single link

We consider the following proportionally-fair congestion control with delayed feedback:

$$\dot{x} = \kappa \left(w - x(t-T)p(x(t-T)) \right). \tag{7.1}$$

To make sure that the user rate x does not become zero, we will assume that the right-hand side is set equal to zero when it is negative and $x = 0$. Recall that the stability condition for the linearized system is given by

$$\kappa(\hat{p} + \hat{x}\hat{p}') < \frac{\pi}{2}. \tag{7.2}$$

Our goal in this section is to derive conditions for the global stability of (7.1) and compare it with the condition (7.2).

Instead of showing global stability first, we begin by showing that the $x(t)$, which is the solution to the delay-differential equation (7.1), is upper and lower bounded around its equilibrium point $\hat{x}$ which is determined by the equation

$$w = \hat{x}p(\hat{x}).$$

Lemma 7.1. *Suppose that $\kappa T \leq \beta$. Given $\varepsilon \geq 0$ and $\varepsilon < w$, there exists $t_0(\varepsilon) < \infty$, such that*

$$l_\varepsilon(\beta) \leq x(t) \leq M_\varepsilon(\beta), \qquad \forall t \geq t_0,$$

where $M_\varepsilon(\beta)$ is the smallest real number satisfying

$$(M_\varepsilon - \beta w)p(M_\varepsilon - \beta w) - w \geq \varepsilon$$

and $l_\varepsilon(\beta) > 0$ is the largest real number satisfying

$$[l_\varepsilon + \beta M_\varepsilon p(M_\varepsilon) - \beta w]p(l_\varepsilon + \beta M_\varepsilon p(M_\varepsilon) - \beta w) - w \leq -\varepsilon.$$

Proof. In this proof, β and ε are assumed to be fixed and so we simply use M and l to denote $M_\varepsilon(\beta)$ and $l_\varepsilon(\beta)$ respectively. We will first show that M is indeed an eventual upper bound. If $x(t) > M$, then

$$x(t - d) \geq M - \kappa w d \geq M - \beta w,$$

which implies that

$$\dot{x}(t) \leq \kappa(w - (M - \beta w)p(M - \beta w))$$
$$\leq -\kappa\varepsilon .$$

So, clearly, there exists $t_1(\varepsilon) < \infty$ such that the trajectory is upper bounded by M for all $t \geq t_1(\varepsilon)$.

Next, we will show the existence of the eventual lower bound. Now, observe the following for $t \geq t_1(\varepsilon) + T$. If $x(t) \leq l$, then

$$x(t - T) \leq l - \kappa T(w - Mp(M)) \leq l + \beta Mp(M) - \beta w$$

which implies that

$$\dot{x}(t) \geq \kappa[w - [l + \beta Mp(M) - \beta w]p(l + \beta Mp(M) - \beta w)]$$
$$\geq \kappa\varepsilon.$$

$\square$

The next step in deriving conditions for global stability is to put the congestion controller (7.1) in the form

$$\dot{y} = -\kappa a(t)y(t - T). \tag{7.3}$$

To do this, we perform the following manipulations on the congestion control equations. Since $w = p(\hat{x})\hat{x}$, we have

$$\dot{x} = \kappa(w - x(t - T)p(x(t - T))$$
$$= \kappa\left(\hat{x}p(\hat{x}) - x(t - d)p(x(t - d))\right)$$
$$= -\kappa a(t)(x(t - d) - \hat{x}),$$

where $a(t)$ is given by

$$a(t) = \frac{x(t - T)p(x(t - T)) - \hat{x}p(\hat{x})}{x(t - T) - \hat{x}}.$$

If the function $p(x)$ is differentiable, then, by the mean-value theorem of calculus, we have

$$a(t) = \frac{d(xp(x))}{dx}\Big|_{x=u_t}$$
$$= p(u_t) + u_t p'(u_t),$$

where u_t is some number between $x(t - T)$ and $\hat{x}$. Defining

$$y(t) = x(t) - \hat{x},$$

we get the required form (7.3). It is useful to note that $a(t)$ is upper-bounded by

$$a_{max}(\beta) = \max_{l_\epsilon(\beta) \leq x \leq M_\epsilon(\beta)} p(x) + xp'(x).$$

If $p(x)$ is not differentiable, but it satisfies the condition

$$|xp(x) - zp(z)| < L|x - z|, \quad \forall x, z \geq 0,$$

for some $L > 0$, then, it is clear that $a(t) < L$. In this case, define $a_{max} = L$.
Now, consider the Lyapunov function

$$V(t) = \max_{t - 2T \leq s \leq t} \frac{y^2(s)}{2}. \tag{7.4}$$

Instead of studying the behavior of a function such as $(y^2/2)$ at a particular time instant t, the key idea in studying the stability of delay-differential equations is to study its behavior over some interval of time as above. This idea is embedded in an extension of Lyapunov theory to delay-differential equations and is presented in a series of results called Razumikhin theorems in [34]. For this reason, we will call the function (7.4) a Lyapunov-Razumikhin function.

We will first provide an informal derivation of the sufficient conditions for global stability. First, define

$$W(y) = \frac{y^2}{2}.$$

Thus, from the definition of $V(y)$ in (7.4), it is clear that

$$V(t) = W(y(\tau)),$$

for some $\tau \in [t - 2T, t]$. If $\tau < t$, then

$$V(t + \delta) \approx V(y(t))$$

for small δ. Thus, when $\tau < t$,

$$\dot{V} = \lim_{\delta \to 0} \frac{V(y(t + \delta)) - V(y(t))}{\delta} = 0.$$

When $\tau = t$, then

$$\begin{aligned}
\dot{V} &= y(t)\dot{y}(t) \\
&= -\kappa a(t)y(t)y(t - T) \\
&= -\kappa a(t)y(t)\left(y(t) - \int_{t-T}^{t} \dot{y}(s)ds \right) \\
&= -\kappa a(t)y^2(t) + \kappa^2 a(t)y(t) \int_{t-T}^{t} a(s)y(s - T)ds.
\end{aligned}$$

Thus, if $\kappa T < \beta$,

$$a(s) \le a_{max}(\beta)$$

ultimately. Let us suppose that we are only considering the system at a time when this upper bound on $a(s)$ holds. Therefore,

$$\begin{aligned}
\dot{V} &= -\kappa a(t)y^2(t) + \kappa^2 a(t)y(t) \int_{t-T}^{t} a(s)y(s - T)ds \\
&\le -\kappa a(t)y^2(t) + \kappa^2 |y(t)|a(t)a_{max} \int_{t-T}^{t} |y(t)|ds,
\end{aligned}$$

where the inequality follows from the fact that $W(t) \ge W(s)$, $\forall s \in [t - 2T, t]$. Thus,

$$\dot{V} \le \kappa a(t)y^2(t)(-1 + \kappa a_{max}(\beta)T),$$

and $\dot{V} < 0$ if

$$\kappa a_{max}(\beta)T < 1.$$

Recalling that we have assumed that $\kappa T < \beta$, the sufficient condition may be rewritten as

$$\kappa T < \max \left(\frac{1}{a_{max}(\beta)}, \beta \right). \tag{7.5}$$

To summarize the discussion so far, we have shown that whenever

$$V(y) = \max_{s \in [t-2T, t]} \frac{y^2(s)}{2},$$

$\dot{V} < 0$. Suppose that the initial condition for (7.3) is of the form

$$x(t) = \phi(t), \qquad t \in [-T, 0],$$

where

$$|\phi(t)| \leq \varepsilon;$$

then it is clear that $|y(t)| < \varepsilon$, $\forall t > 0$. Thus, the system is locally, asymptotically stable.

To get a condition for global stability, we need the following Razumikhin theorem [34].

Theorem 7.2. *Consider the differential equation*

$$\dot{y} = -a(t)y(t - T),$$

along with an initial trajectory specified in the interval $[-T, 0]$. *Choose a function* $W(y) \geq 0$ *such that* $W(0) = 0$ *and* $W(y) \rightarrow \infty$ *as* $y \rightarrow \infty$. *If there exists* $q > 1$ *such that*

$$\dot{W} \leq 0,$$

whenever

$$qW(t) \geq \max_{t-T \leq s \leq t} W(s),$$

then the system is globally, asymptotically stable.   □

The above theorem is a special case of Theorem 4.2 in [34]. Comparing the above theorem to the discussion prior to the theorem, we note that the theorem requires $W(t)$ to be decreasing even when $W(t)$ is smaller than $\max_{t-T \leq s \leq t} W(s)$. Specifically, the theorem requires that $W(t)$ should decrease when it is greater than or equal to $\max_{t-T \leq s \leq t} W(s)$ by a factor $1/q$. Assuming that

$$qW(t) \geq \max_{t-T \leq s \leq t} W(s),$$

we have

$$\dot{W} \leq -\kappa a(t)y^2(t) + \kappa^2 a(t)y(t) \int_{t-2T}^{t} a(s)y(s)ds$$

$$\leq -\kappa a(t)y^2(t) + \sqrt{q}\kappa^2|y(t)|a(t)a_{max} \int_{t-2T}^{t-T} |y(t)|ds$$

$$\leq \kappa a(t)y^2(t)(1 - \sqrt{q}\kappa a(t)T).$$

Thus, if

$$\kappa \left(\max_{t-T \leq s \leq t} a(t) \right) T < 1,$$

then we can find $q > 1$ such that $\dot{W} < 0$. In other words, (7.5) is also a condition for global stability. We summarize our discussion so far in this chapter in the following theorem.

Theorem 7.3. *If the condition (7.5) is satisfied, then the system (7.1) is globally, asymptotically stable.* □

We note that the condition for global stability in (7.5) can be relaxed to make the right-hand side of (7.5) equal to 3/2, instead of 1. When the condition (7.5) is violated, it is sometimes possible to compute a bound on the initial conditions for which the solution to (7.1) is asymptotically stable. We refer the interested reader to [19] for further details.

In the following two examples, we present two applications of Theorem 7.3.

Example 7.4. Suppose that the congestion signals are generated by a virtual queue with a finite buffer. Using the approximation

$$p(x) = \left(\frac{x-C}{x}\right)^+ ,$$

where C is the capacity of the virtual queue, we get $p(x)+xp'(x) \equiv 1, \forall x \geq C$. It can also be shown from Lemma 7.1 that $M(\beta) = x^*+\beta w$ is an upper bound for this system. It is easy to see that

$$\frac{|xp(x) - x^*p(x^*)|}{|x - x^*|} \leq 1$$

for all x. Using Theorem 7.3, a sufficient condition for stability is $\kappa d < 1$ which is slightly weaker than the condition $\kappa d < \pi/2$ stated in [49]. □

Example 7.5. Recall that REM marks a packet with probability $(1-\exp(-\theta b))$ if the packet sees b other packets in the virtual queue. If the queue length hits zero many times during an RTT, using a reflected Brownian motion approximation, it has been suggested in [49] that REM can be viewed as a mechanism with the marking function

$$p(x) = \frac{\theta\sigma^2 x}{\theta\sigma^2 x + 2(C - x)}. \tag{7.6}$$

The reflected Brownian motion approximation will be introduced in a later chapter. For now, we can simply consider the $p(x)$ in (7.6) as a marking function in itself, instead of as an approximation to REM. Here σ^2 denotes the variability of the traffic at the packet level and C is the capacity of the virtual queue. We assume $\theta\sigma^2 = 0.5$.

Let the capacity of the virtual queue be 1 Mbps and the round trip delay be 40 ms. Let one round trip time be the unit of time. If the packet sizes are 1000 bytes each, then the virtual queue capacity can be equivalently expressed as five packets per time unit. Let the increase parameter w be 1, so that the controller is

$$\dot{x} = \kappa[1 - x(t - 1)p(x(t - 1))],$$

where

$$p(x) = \frac{x}{20 - 3x} \ .$$

The equilibrium rate can be found by solving $xp(x) = 1$ yielding $x^* = 3.22$. Choose $\beta = 0.83$ and $\varepsilon = 0.01$ in Lemma 7.1. The eventual upper and lower bounds can be calculated as $M = 4.06$ and $l = 2.29$, respectively. The sufficient condition for global stability is given by $\kappa d < \min[0.83, 0.5427] = 0.5427$. Different choices for β will yield different conditions on κd for global stability. We also note that our global stability condition is weaker than the condition for the linearized version to be stable, which is $\kappa d < 1.729$.   □

We conclude this chapter by noting that similar single link results have also been obtained for the dual controller in [95] and Razumikhin's theorem has been applied to study the region of attraction of Jacobson's TCP congestion control algorithm on a single link in [36]. Extensions to the general topology case can be found in [1, 22]; however, these results only establish stability when the feedback delays are extremely small. In other words, the bounds on delays for the general topology network are extremely restrictive compared to the linear analysis. Such results were also obtained for the dual algorithm by analyzing a discrete-time version of the algorithm in [69].

8

Stochastic Models and their Deterministic Limits

In the previous chapters, we have used deterministic models to derive congestion control schemes and AQM schemes that are fair and achieve high utilization. However, in the real Internet, there are many sources of randomness:

- unresponsive flows which do not respond to congestion indication,
- the probabilistic nature of packet marking by an AQM scheme,
- asynchronous updates among sources,
- the inability to precisely model window flow control mechanism, and
- the initial ramp-up phase (for example, *slow start* in TCP flow control) of the congestion control mechanism.

In this chapter, we will start with stochastic models of a single link accessed by congestion controlled sources and show that, when the the number of users in the network is large, one arrives at deterministic equations that characterize the system behavior. This is analogous to the well-known law of large numbers for random variables [33]. In addition, we will also derive stochastic models to describe the variability around the deterministic limit, much like the central limit theorem for random variables [33]. Our goal in this chapter is to model the interaction between congestion control at the sources and congestion indication mechanisms at the router. This is different from the models in [81, 3, 6] which assume that the marking or drop probability at the link is independent of the arrival rate at the link.

For simplicity of exposition, we will only consider delay-free models in this section. Further, we will restrict our attention to a single node, shared by N sources with identical RTTs. Recall that the proportionally-fair congestion controller is given by

$$\dot{x}_r = \kappa \left(w - x_r p_N (\sum_{r=1}^{N} x_r) \right),$$

where $p_N(\cdot)$ is the marking probability at the link. Defining

$$x := \frac{1}{N} \sum_{r=1}^{N} x_r,$$

we get

$$\dot{x} = \kappa \left(w - xp(x) \right), \tag{8.1}$$

where we have defined $p(x) = p_N(Nx)$. The discrete-time analog of (8.1) is given by

$$x(k+1) = x(k) + \kappa \left(w - x(k)p(x(k)) \right), \tag{8.2}$$

assuming that the discrete time slot is of duration 1 time unit. Our goal in this section is to start with a stochastic model of congestion control flows and show that, under an appropriate scaling, one obtains (8.2) as the limit.

8.1 Deterministic limit for proportionally-fair controllers

Consider N identical congestion-controlled sources accessing a link. The dynamics of the rth source is described by

$$x_r^{(N)}(k+1) = x_r^{(N)}(k) + \kappa(w - M_r(k)), \tag{8.3}$$

where $x_r^{(N)}(k+1)$ is the transmission rate (measured in packets per unit time) of source r at time slot k and $M_r(k)$ is a random variable denoting the number of marks received by the rth source at time instant k. If the congestion control is window-based, and if one discrete-time slot is equal to the RTT, then $x_r^{(N)}(k)$ can be interpreted as the window size of source r at time instant k. The window size does not have to be an integer, and therefore, the number of packets transmitted in time slot k would be $\lfloor x_r^{(N)}(k) \rfloor$. Further, the various sources need not be perfectly time-synchronized and as described at the beginning of the chapter, there are several other network phenomena that are impossible to model precisely. Therefore, we make the simplifying assumption, that while $x_r^{(N)}(k)$ is the intended rate at time slot k, the actual number of packets transmitted in slot k by source r is a Poisson random variable with mean $x_r^{(N)}(k)$. We will denote a Poisson random variable with mean λ by $Poi(\lambda)$. We will assume that the number of packets generated by the different sources at different time slots are independent, i.e., $\{Poi(x_r^{(N)}(k))\}$, are independent.

The marking process at the link marks each packet with probability $\xi_N(k)$ at time instant k. The probability $\xi_N(k)$ is assumed to be a function of $y_N(k)$, the total arrival rate at the link. Specifically, let $\xi_N(k) = p_N(y_N(k))$, where $p(y)$ is assumed to be of the form

$$p_N(y) = 1 - exp\left(-\frac{\gamma}{N} y \right).$$

Note that $p_N(y)$ is an increasing function of y, reminiscent of REM. However, here we are assuming that the marking probability is simply chosen to be a function of arrival rate, rather than the queue length at the link.

Next, define the average source rate $x^{(N)}(k)$ to be

$$x^{(N)}(k) = \frac{1}{N} \sum_{r=1}^{N} x_r^{(N)}(k).$$

The dynamics of the average source rate is given by

$$x^{(N)}(k+1) = x^{(N)}(k) + \kappa \left(w - \frac{1}{N} \sum_{r=1}^{N} M_r(k) \right). \tag{8.4}$$

In what follows, we will show that, under appropriate conditions,

$$\lim_{N \to \infty} \frac{1}{N} \sum_{r=1}^{N} M_r(k) = x(k)(1 - e^{-\gamma x(k)}) \qquad \text{in probability,}$$

and therefore,

$$x^{(N)}(k+1) \to x(k+1),$$

where $x(k+1)$ satisfies (8.2) with $p(x) = 1 - e^{-\gamma x}$. Our model is a combination of the models in [90, 91]. The large number of sources scaling and the variability in the arrival process at the link were introduced in [90], but the deterministic limit with probabilistic marking was first studied in [91].

We will first provide a heuristic derivation to show why this result may be expected. The following properties of Poisson random variables will be useful.

1. The mean and variance of a Poisson random variable are given by

$$E\left(Poi(\lambda)\right) = Var\left(Poi(\lambda)\right) = \lambda.$$

2. If X_1 and X_2 are independent Poisson random variables with means λ_1 and λ_2, respectively, then $X_1 + X_2$ is a Poisson random variable with mean $\lambda_1 + \lambda_2$.

3. Let X be a Poisson random variable with mean λ, denoting the number of packet arrivals in a unit time at a link. Now, suppose that each packet is marked independently with probability p, and Y is the random variable denoting the number of marked packets. Then, Y is a Poisson random variable with mean $p\lambda$.

Now, suppose that a law of large numbers holds for $x^{(N)}(0)$, i.e.,

$$x^{(N)}(0) \to x(0),$$

in an appropriate sense, for some $x(0)$. Next, note that

$$E\left(\sum_{r=1}^{N} M_r(1)\right) = E\left[E\left(\sum_{r=1}^{N} M_r(1) \,|\, \xi_N(0), \{x^{(N)}(0)\}\right)\right].$$

Further, given $\xi_N(0)$ and $\{x^{(N)}(0)\}$,

$$M_r(1) \sim Poi\left(x_r^{(N)}(0)\xi_N(0)\right),$$

and since $\{M_r(1)\}$ are independent,

$$\sum_r M_r(1) \sim Poi\left(Nx^{(N)}(0)\xi_N(0)\right).$$

Thus,

$$E\left(\sum_{r=1}^{N} M_r(1) \,|\, \xi_N(0), \{x^{(N)}(0)\}\right) = Nx^{(N)}(0)\xi_N(0)$$

and

$$E\left(\sum_{r=1}^{N} M_r(1)\right) = E\left(Nx^{(N)}(0)\xi_N(0)\right) = E\left[Nx^{(N)}(0)\xi_N(0) \,|\, \{x_r^{(N)}(0)\}\right].$$

Next, note that, given $\{x_r^{(N)}(0)\}$,

$$y_N(0) = \sum_{r=1}^{N} Poi\left(x_r(0)\right) = Poi\left(Nx^{(N)}(0)\right).$$

Thus,

$$
\begin{aligned}
E\left(\sum_{r=1}^{N} M_r(1)\right) &= E\left[Nx^{(N)}(0)E\left(1 - e^{-\gamma y_N(0)/N} \,|\, \{x_r^{(N)}(0)\}\right)\right] \\
&= E\left[Nx^{(N)}(0)\sum_{i=1}^{\infty}(1 - e^{-\gamma i/N})\frac{(Nx^{(N)}(0))^i e^{-Nx^{(N)}(0)}}{i!}\right] \\
&= E\left[Nx^{(N)}(0)e^{-Nx^{(N)}(0)}\left(1 - \sum_{i=1}^{\infty}\frac{(Nx^{(N)}(0)e^{-\gamma/N})^i}{i!}\right)\right] \\
&= E\left[Nx^{(N)}(0)e^{-Nx^{(N)}(0)}\right. \hspace{3cm} (8.5) \\
&\qquad \left. \times\left(e^{Nx^{(N)}(0)} - exp(Nx^{(N)}(0)e^{-\gamma/N})\right)\right] \\
&= E\left[Nx^{(N)}(0)\left(1 - exp(-Nx^{(N)}(0)(1 - e^{-\gamma/N}))\right)\right]. \quad (8.6)
\end{aligned}
$$

Since

$$x^{(N)}(0) \to x(0),$$

under appropriate conditions, it seems reasonable to expect that

$$\frac{1}{N} E \left(\sum_{r=1}^{N} M_r(1) \right) \to x(0) \left(1 - e^{-\gamma x(0)} \right).$$

Further, since we are dealing with the average of a large number of random variables, it seems reasonable to expect the following law-of-large-numbers result:

$$\frac{1}{N} \sum_{r=1}^{N} M_r(1) \to x(0) \left(1 - e^{-\gamma x(0)} \right)$$

in an appropriate sense. If the above limit is correct, then

$$x^{(N)}(1) \to x(1) = x(0) + \kappa \left(w - x(0)(1 - e^{-\gamma x(0)}) \right).$$

Then, by induction, one can expect (8.2) to hold. In the rest of this section, the above discussion is made precise.

We start with the following lemma.

Lemma 8.1. *The following identities hold:*

$$E \left(\sum_{r=0}^{N} M_r^{(N)}(k+1) \right) = E \left\{ N x^{(N)}(k) \left(1 - exp \left[-N x^{(N)}(k)(1 - e^{-\gamma/N}) \right] \right) \right\}$$

$$E \left\{ \left(\sum_{r=0}^{N} M_r^{(N)}(k+1) \right)^2 \right\}$$

$$= E \left\{ N x^{(N)}(k) \left(1 - exp(-N x^{(N)}(k)(1 - e^{-\gamma/N})) \right) \right.$$

$$+ N^2 (x^{(N)}(k))^2 \left(1 - 2exp(-N x^{(N)}(k)(1 - e^{-\gamma/N})) \right.$$

$$\left. + exp(-N x^{(N)}(k)(1 - e^{-2\gamma/N})) \right) \right\}.$$

Proof. The first identity was shown for the case $k = 0$, using properties of Poisson random variables, in the discussion prior to the lemma. The derivation for $k > 0$ is identical to the case $k = 0$. The second identity can be shown in a similar manner as follows:

$$E\left\{\left(\sum_{r=0}^{N} M_r^{(N)}(k+1)\right)^2\right\} = E\left\{\left(\sum_{r=0}^{N} M_r^{(N)}(k)\right)^2 \mid \xi_N(k), \{x_r^{(N)}(k)\}\right\}$$

$$= E\left\{E\left(Poi(Nx^{(N)}(k)\xi_N(k))\right)\right\}$$

$$= E\left\{Nx^{(N)}(k)\xi_N(k) + \left(Nx^{(N)}(k)\xi_N(k)\right)^2\right\}$$

$$= E\left\{Nx^{(N)}(k)(1 - e^{-\gamma y_N(k)/N})\right.$$

$$\left. + \left(Nx^{(N)}(k)(1 - e^{-\gamma y_N(k)/N})\right)^2\right\}.$$

Now, the rest of the proof follows along the lines of the derivation in (8.6). □

Now we are ready to state and prove the main result of this section.

Theorem 8.2. *Suppose that*

$$\lim_{N\to\infty} E\left((x^{(N)}(0) - x(0))^2\right) = 0.$$

Then, given any $\varepsilon > 0$,

$$\lim_{N\to\infty} P\left(|x^{(N)}(k) - x(k)| > \varepsilon\right) = 0, \qquad \forall k \geq 0,$$

where $\{x(k)\}$ is defined by the recursion (8.2), with $p(x) = 1 - e^{-\gamma x}$.

Proof. We will show that

$$\lim_{N\to\infty} E\left((x^{(N)}(k) - x(k))^2\right) = 0, \qquad \forall k \geq 0. \tag{8.7}$$

Then, from the Chebyshev-Markov inequality, i.e.,

$$P\left(|x^{(N)}(k) - x(k)| > \varepsilon\right) \leq \frac{1}{\varepsilon^2} E\left((x^{(N)}(k) - x(k))^2\right),$$

the desired result follows.

We prove (8.7) by induction. It is true for $k = 0$ by the assumption in the statement of the theorem. Assuming that it is true for some $k > 0$, we will prove that it holds for $(k+1)$.

$$E\left((x^{(N)}(k+1) - x(k+1))^2\right)$$

$$= E\left[\left(x^{(N)}(k) + \kappa(w - \tfrac{1}{N}\sum_{r=1}^{N} M_r^{(N)}(k)) - x(k) - \kappa(w - x(k)p(x(k)))\right)^2\right]$$

$$= E\left[\left(x^{(N)}(k) - x(k) + \tfrac{1}{N}\sum_{r=1}^{N} M_r^{(N)}(k) - x(k)p(x(k))\right)^2\right]$$

$$= E\left(x^{(N)}(k) - x(k)\right)^2 + E\left(\frac{1}{N}\sum_{r=1}^{N} M_r^{(N)}(k) - x(k)p(x(k))\right)^2$$

$$+ 2E\left[\left(x^{(N)}(k) - x(k)\right)\left(\frac{1}{N}\sum_{r=1}^{N} M_r^{(N)}(k) - x(k)p(x(k))\right)\right].$$

$$(8.8)$$

By the induction hypothesis,

$$\lim_{N\to\infty} E\left(x^{(N)}(k) - x(k)\right)^2 = 0.$$

We will show that

$$\lim_{N\to\infty} E\left(\frac{1}{N}\sum_{r=1}^{N} M_r^{(N)}(k) - x(k)p(x(k))\right)^2 = 0. \qquad (8.9)$$

The remaining term in (8.8) can be shown to be equal to zero in a similar manner.

To show (8.9), we note that

$$E\left(\frac{1}{N}\sum_{r=1}^{N} M_r^{(N)}(k) - x(k)p(x(k))\right)^2$$

$$= E\left(\frac{1}{N}\sum_{r=1}^{N} M_r^{(N)}(k)\right)^2 - 2E\left[\left(\frac{1}{N}\sum_{r=1}^{N} M_r^{(N)}(k)\right)(x(k)p(x(k)))\right]$$

$$+ (x(k)p(x(k)))^2$$

$$= E\left[(x(k)p(x(k)))^2 - \left(\frac{1}{N}\sum_{r=1}^{N} M_r^{(N)}(k)\right)(x(k)p(x(k)))\right] \qquad (8.10)$$

$$+ E\left[\left(\frac{1}{N}\sum_{r=1}^{N} M_r^{(N)}(k)\right)^2 - \left(\frac{1}{N}\sum_{r=1}^{N} M_r^{(N)}(k)\right)(x(k)p(x(k)))\right] \quad (8.11)$$

We will first show that (8.10) goes to 0 as $N \to \infty$. From Lemma 8.1,

$$\left| E\left[x(k)p(x(k)) - \left(\frac{1}{N}\sum_{r=1}^{N} M_r^{(N)}(k) \right) \right] \right|$$

$$= \left| E\left[x(k)p(x(k)) - x^{(N)}(k)\left(1 - exp\left[-Nx^{(N)}(k)(1 - e^{-\gamma/N}) \right] \right) \right] \right|$$

$$\le E\left(\left| (x(k) - x^{(N)}(k))p(x(k)) \right| \right) + E\left(\left| x^{(N)}(k)\left(p(x(k)) - p(x^{(N)}(k)) \right) \right| \right)$$

$$+ E\left(\left| x^{(N)}(k)\left(p(x^{(N)}(k)) - 1 + exp\left[-Nx^{(N)}(k)(1 - e^{-\gamma/N}) \right] \right) \right| \right)$$

It is easy to see that each of the three terms on the right-hand side of the above inequality go to zero. Thus, (8.10) goes to 0 as $N \to \infty$.

Next we show that (8.11) goes to 0 as $N \to \infty$. From Lemma 8.1,

$$E\left[\left(\frac{1}{N}\sum_{r=1}^{N} M_r^{(N)}(k) \right)^2 - \left(\frac{1}{N}\sum_{r=1}^{N} M_r^{(N)}(k) \right)(x(k)p(x(k))) \right]$$

$$= E\left\{ \frac{1}{N}x^{(N)}(k)\left(1 - exp(-Nx^{(N)}(k)(1 - e^{-\gamma/N})) \right) \right.$$

$$+ (x^{(N)}(k))^2 q_N(x^N(k))$$

$$\left. - x(k)x^{(N)}(k)p(x(k))\left(1 - exp\left[-Nx^{(N)}(k)(1 - e^{-\gamma/N}) \right] \right) \right\},$$

where

$$q_N(x) := 1 - 2exp\left(-Nx(1 - e^{-\gamma/N}) \right) + exp\left(-Nx^{(N)}(k)(1 - e^{-2\gamma/N}) \right).$$

Now,

$$\left| E\left[\left(\frac{1}{N}\sum_{r=1}^{N} M_r^{(N)}(k) \right)^2 - \left(\frac{1}{N}\sum_{r=1}^{N} M_r^{(N)}(k) \right)(x(k)p(x(k))) \right] \right|$$

$$\le E\left[\left| \frac{1}{N}x^{(N)}(k)\left(1 - exp(-Nx^{(N)}(k)(1 - e^{-\gamma/N})) \right) \right| \right]$$

$$+ E\left[\left| \left((x^{(N)}(k))^2 - x(k)x^{(N)}(k) \right) q_N(x^N(k)) \right| \right]$$

$$+ E\left[\left| x(k)x^{(N)}(k)\left(p(x(k))\left(1 - exp\left[-Nx^{(N)}(k)(1 - e^{-\gamma/N}) \right] \right) \right. \right. \right.$$

$$\left. \left. \left. - q_N(x^{(N)}(k)) \right) \right| \right].$$

Again, it is easy to see that each term on the right-hand side of the above inequality goes to zero, thus completing the proof of the theorem. $\square$

Remark 8.3. Suppose that the router uses a different marking function, say, $p_N(y) = (y/Nc)^B$. Then, it is not true, in general, that the limiting deterministic dynamics for the average source rate would be of the form

$$x(k+1) = x(k) + \kappa \left(w - x(k)(x(k)/c)^B \right).$$

The reason for this can be understood by looking at the derivation of (8.6). It was somewhat fortuitous that the limiting marking function in (8.6) had the same form as the original marking function. Using the marking function $(y/Nc)^B$, the resulting expression for

$$\frac{1}{N} \sum_r M_r^{(N)}(k)$$

becomes quite a bit more complicated than in (8.6). In general, the limit would be of the form

$$x(k+1) = x(k) + \kappa \left(w - x(k)p(x(k)) \right)$$

for some marking function $p(x)$ which may or may not be identical to the actual marking function used by the router. $\qquad\qquad \Box$

8.2 Individual source dynamics

In the previous section, we showed that the average source rate obeys a deterministic difference equation. In this section, we will study the dynamics of an individual source. While the average source rate obeys a law of large numbers, one would expect the individual source rates to exhibit random fluctuations. Our goal in this section is to provide models for the random nature of the individual source rate. Specifically, we are interested in computing the variance of the individual source rate. Our derivations in this section will be heuristic without providing formal proofs of our results.

Before we consider the behavior of individual source rates, let us first consider the fluctuations of the average rate $x^{(N)}(k)$ about its limit $x(k)$. Since $x^{(N)}(k)$ converges in probability to $x(k)$, it is reasonable to expect a central limit theorem for

$$\sqrt{N} \left(x^{(N)}(k) - x(k) \right).$$

From (8.8), it can be verified that $E(x^{(N)}(k+1) - x(k+1))^2$ is $O(1/N)$. Thus, under appropriate conditions, we can show that a central limit theorem type result holds. However, we do not prove that here. Rather, we simply use the fact that $(x^{(N)}(k) - x(k))$ is $O(1/\sqrt{N})$ with high probability and thus, in the calculation of the variance of the individual source rate, we will replace $x^{(N)}(k)$ by the deterministic quantity $x(k)$. Thus, for large N, the dynamics of source r's rate are given by

$$x_r(k+1) = x_r(k) + \kappa \left(w - Poi(x_r(k)(1 - e^{-\gamma x(k)})) \right), \qquad (8.12)$$

where $x(k)$ is given by (8.2) with $p(x) = 1 - e^{-\gamma x}$.

Denoting the expected value of $x_r(k)$ by $\bar{x}_r(k)$, and taking expectations on both sides of (8.12), we get

$$\bar{x}_r(k+1) = \bar{x}_r(k) + \kappa \left(w - \bar{x}_r(k)(1 - e^{-\gamma x(k)}) \right). \tag{8.13}$$

Next, from (8.12) and (8.13), we have

$$x_r(k+1) - \bar{x}_r(k+1) = x(k) - \bar{x}_r(k)$$
$$- \kappa \left(Poi(x_r(k)(1 - e^{-\gamma x(k)})) - \bar{x}_r(k)(1 - e^{-\gamma x(k)}) \right).$$

Letting $\sigma_k := E(x(k) - \bar{x}_r(k))^2$, we obtain

$$\sigma_{k+1}^2 = \sigma_k^2 + \kappa^2 E \left(Poi(x_r(k)(1 - e^{-\gamma x(k)}) - \bar{x}_r(k)(1 - e^{-\gamma x(k)}) \right)^2$$
$$- 2\kappa E \left[(x(k) - \bar{x}_r(k)) \left(Poi(x_r(k)(1 - e^{-\gamma x(k)})) \right. \right. \tag{8.14}$$
$$\left. \left. - \bar{x}_r(k)(1 - e^{-\gamma x(k)}) \right) \right]$$

$$= \left(1 - \kappa(1 - e^{-\gamma x(k)}) \right)^2 \sigma_k^2 + \kappa^2 \bar{x}_r(k) \left(1 - e^{-\gamma x(k)} \right) \tag{8.15}$$

If (8.13) is stable, the steady-state mean of any source r is given by the solution to equation

$$w = \hat{\bar{x}}_r \left(1 - e^{-\gamma \hat{x}} \right).$$

If (8.15) is stable, the steady-state variance of any source r is given by

$$\hat{\sigma}_r^2 = \frac{\kappa^2 \hat{\bar{x}}_r \left(1 - e^{-\gamma \hat{x}} \right)}{1 - \left(1 - \kappa \left(1 - e^{-\gamma \hat{x}} \right) \right)^2},$$

where $\hat{x}$ is the equilibrium value of (8.2) and is given by

$$w = \hat{x}_r \left(1 - e^{-\gamma \hat{x}} \right).$$

Next, let us derive conditions for the stability of (8.13) and (8.15). To do this, we have to first study the stability of (8.2). Linearizing (8.2), we obtain

$$\delta x(k+1) = \delta x(k) - \kappa(\hat{p} + \hat{x}\hat{p}')\delta x(k).$$

Thus, the condition for local stability of (8.2) is

$$\kappa(\hat{p} + \hat{x}\hat{p}') < 2. \tag{8.16}$$

If the RTT is equal to 1 time unit, this condition is very similar to the stability condition in Theorem 6.1. This is to be expected since the model in this section is the discrete-time analog of the model for which Theorem 6.1 holds.

We will use the following lemma to derive the stability conditions for (8.13) and (8.15).

Lemma 8.4. *Consider the discrete-time system*

$$x(k+1) = \rho_k x(k) + b.$$

If $\rho_k^2 \to \rho^2 < 1$ as $k \to \infty$, then, starting from any initial condition,

$$\lim_{k \to \infty} x(k) = \frac{b}{1-\rho}.$$

Proof. Let

$$x_\infty := \frac{b}{1-\rho},$$

and define

$$y(k) = x(k) - x_\infty.$$

Then, we have

$$y(k+1) = \rho_k y(k) + \varepsilon_k,$$

where

$$\varepsilon_k = b\left(\frac{\rho_k - \rho}{1-\rho}\right).$$

Consider the Lyapunov function $V_k = y_k^2$. We have

$$
\begin{aligned}
V_{k+1} - V_k &= y^2(k+1) - y^2(k) \\
&= (\rho_k^2 - 1)y_k^2 + \varepsilon_k^2 + 2\rho_k y_k \varepsilon_k \\
&= (\rho_k^2 - 1)\left(y(k) - \frac{\rho_k \varepsilon_k}{1 - \rho_k^2}\right) - \frac{\varepsilon_k^2}{(1 - \rho_k^2)^2} \\
&< 0,
\end{aligned}
$$

for large enough k, since $\varepsilon_k \to 0$ and $\rho_k^2 \to \rho < 1$ as $k \to \infty$. Thus, the result follows from a discrete-time version of Lyapunov's theorem given in Appendix 3.10.   □

Applying the above lemma shows that the stability condition for (8.13) and (8.15) is

$$\kappa(1 - e^{-\gamma \hat{x}}) < 2,$$

which is clearly satisfied if (8.16) is satisfied. Thus, the stability of the mean-rate dynamics (8.2) ensures that the mean and variance of the transmission rates of individual sources converge.

The expressions for the steady-state mean and variance of a single source point out an interesting fact. Suppose that the discrete-time unit is a multiple of the RTT. Let the discrete time unit be equal to f times each source's RTT T. Then, the corresponding expressions for σ_2^2 would be

$$\hat{\sigma}_r^2 = \frac{\kappa^2 f T^2 \hat{\bar{x}}_r \left(1 - e^{-\gamma \hat{x}}\right)}{1 - \left(1 - \kappa f T \left(1 - e^{-\gamma \hat{x}}\right)\right)^2}.$$

For a fixed duration of the discrete time slot, f is larger when T is smaller. The results in Chapter 6 suggest that $\kappa T(\hat{p} + \hat{x}\hat{p}')$ should be less than $\pi/2$. It is easy to see that

$$p + xp'(x)$$

is upper-bounded for all $x \geq 0$ when $p(x) = 1 - e^{-\gamma x}$. Let us suppose that this upper bound is L. If κ is chosen to be $1/LT$, then the stability condition is satisfied. However, this results in a larger variance for sources with smaller RTT (since f is larger for such sources). Therefore, it seems more appropriate to choose $\kappa = 1/LT$ when T is larger than a threshold T_{max} and choose κ equal to some constant κ^* when $T \leq T_{max}$, assuming $\kappa LT_{max} < \pi/2$. This simply states the intuitive fact that when κ is large, a rapid increase in the source rate is possible, thus potentially leading to large fluctuations. Thus, a more sensible choice is to choose κ small to control the variance, even though the deterministic stability condition allows for large κ when T is small. This observation was made by Kelly in [50], based on variance calculations by Ott in [80].

8.3 Price feedback

Instead of probabilistic feedback, suppose that the link price is directly fed back to the sources. This would require more than one bit per packet to convey the price information. Such schemes were originally proposed for ATM networks, where a multiple-bit field called the ER (for explicit rate) field was used by the routers to convey to each source the rate at which it should send data into the network (see [8, 4, 44, 38] and references within). More recently, such explicit feedback schemes have been suggested for the Internet in [46]. In this section, we will consider the effect of explicit feedback on the variance of a single flow.

When explicit price feedback is available, the dynamics of source r are given by

$$x_r^{(N)}(k+1) = x_r^{(N)}(k) + \kappa \left(w - Poi(x_r^{(N)}(k))(1 - e^{-\gamma y_N(k))/N}) \right). \quad (8.17)$$

As in the case of probabilistic feedback, it can be shown that the dynamics of

$$x^{(N)}(k) = \frac{1}{N} x_r^{(N)}(k)$$

converge to (8.2) in the limit as $N \to \infty$. For large N, the dynamics of source r is then given by

$$x_r(k+1) = x_r(k) + \kappa \left(w - Poi(x_r(k))(1 - e^{-\gamma x(k)}) \right).$$

Thus, the expectation of x_r has the dynamics

$$\bar{x}_r(k+1) = \bar{x}_r(k) + \kappa\left(w - \bar{x}_r(k)(1 - e^{-\gamma x(k)})\right).$$

Defining

$$\sigma_k^2 = E(x_r(k) - \bar{x}_r(k))^2,$$

we get

$$\sigma_{k+1}^2 = (1 - \kappa p(x(k)))\sigma_k^2 + \kappa^2 p^2(x(k))\bar{x}_r(k).$$

Thus, the steady-state variance is given by

$$\sigma_\infty = \frac{\kappa p^2(x_\infty)x_\infty}{(1 - \kappa p(x_\infty))^2}.$$

Comparing this to the steady-state variance of a single flow in the case of probabilistic feedback, the variance is smaller by a factor equal to $p(x_\infty)$, or equivalently, the standard deviation is smaller by a factor of $\sqrt{p(x_\infty)}$.

In Figures 8.1 through 8.4, we present simulation results which qualitatively illustrate the analytical results presented in this chapter. The source parameters are taken to be $w = 10$, $\kappa = 1$, and the REM parameter is taken to be $\gamma = 0.006$. For these parameter values, the equilibrium rate per user turns out to be 43.5. These figures show that, under both price feedback and probabilistic marking, the average rate seen by the users has much smaller standard deviation compared to the standard deviation of an individual user. Further, probabilistic feedback leads to a larger standard deviation in the rate seen by an individual user as compared to price feedback.

8.4 Queue-length-based marking

Consider N proportionally-fair primal-dual controllers accessing a single link of capacity Nc. The dynamics of the rth source are given by

$$\dot{x}_r = \kappa(w - x_r p_N(\hat{b})), \tag{8.18}$$

where $\hat{b}$ is the queue length at the link. The dynamics of $\hat{b}$ are given by

$$\dot{\hat{b}} = \left[\sum_{i=1}^{N} x_i - Nc\right]_{\hat{b}}^{+}. \tag{8.19}$$

Define $b = \hat{b}/N$ and suppose that $p_N(\hat{b})) = p(b)$. Then $x := (\sum_i x_i)/N$ has the dynamics given by

$$\dot{x} = \kappa(w - xp(b)), \tag{8.20}$$

$$\dot{b} = [x - c]_b^+. \tag{8.21}$$

Examples of functions $p_N(\hat{b})$ and $p(b)$ such that $p_N(\hat{b})) = p(b)$ are given by

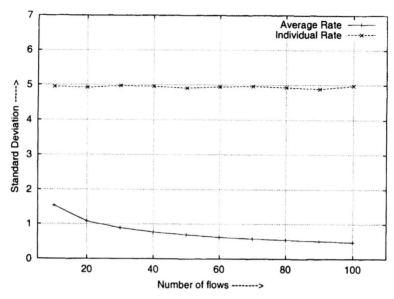

Fig. 8.1. Plot of number of flows versus standard deviation with probabilistic marking

$$p_N(\hat{b})) = 1 - e^{-\gamma\hat{b}/N}, \qquad p(b) = 1 - e^{-\gamma b}.$$

Note that the above marking function is used in the REM algorithm introduced in Chapter 3.

The discrete-time versions of (8.20)and (8.21) are given by

$$x_r(k+1) = x_r(k) + \kappa(w - x_r(k)p_N(b(k))), \tag{8.22}$$

and

$$b(k+1) = b(k) + [x - c]_b^+, \tag{8.23}$$

respectively. As before, we have assumed that the discrete time-step is one time unit long and it is also equal to the common RTT of all the sources.

The stochastic models of (8.18) and (8.19) are given by

$$x_r^{(N)}(k+1) = x_r^{(N)}(k) + \kappa(w - M_r^{(N)}(k+1)), \tag{8.24}$$

and

$$\tilde{b}^{(N)}(k+1) = \tilde{b}^{(N)}(k) + \left[\sum_{i=1}^{N} x_i^{(N)}(k) - c\right]_{\tilde{b}^N(k)}^+, \tag{8.25}$$

respectively. Analogous to the stochastic model in Section 8.1, assume that each packet is marked with probability $\xi_N(k)$, where $\xi_N(k) = 1 - e^{-\gamma\tilde{b}^{(N)}(k)/N}$. Letting $p(b) = 1 - e^{-\gamma b}$ and mimicking the approach in Section 8.1, we can show that

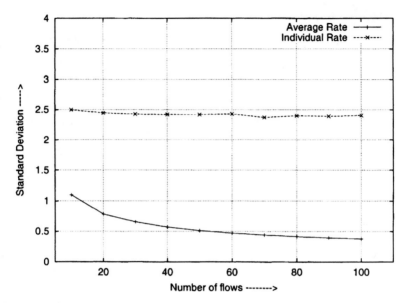

Fig. 8.2. Plot of number of flows versus standard deviation with price feedback

$$x^{(N)}(k) := \frac{1}{N} \sum_{r=1}^{N} x_r^{(N)}(k)$$

and

$$b^{(N)}(k) := \frac{b^{(N)}(k)}{N}$$

converge in probability to (8.22) and (8.23) as $N \to \infty$. The details involve a combination of the proof in Section 8.1 and the results in [91] where a similar result has been proved for a related model.

In general, just because the marking is based on queue length does not necessarily mean that the queue length will explicitly appear in the deterministic limit model. It depends on the scaling used in the AQM scheme and this is discussed further in [20].

8.5 TCP-type congestion controllers

Consider the following stochastic model for a TCP-type controller:

$$x_r^{(N)}(k+1) = x_r^{(N)}(k) + \kappa(w - x_r^{(N)} M_r^{(N)}(k)), \tag{8.26}$$

with a rate-based marking function model for $M_r^{(N)}(k+1))$ as in Section 8.1. Recall that (8.26) corresponds to a source with utility function $-w/x_r$. As before, defining

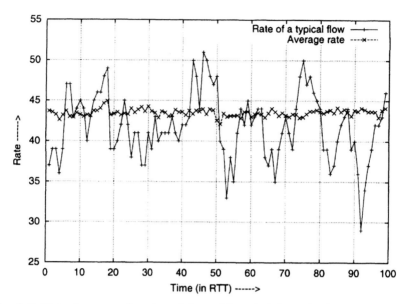

Fig. 8.3. Plot of rate as a function of time with probabilistic feedback. The number of source is 100

$$x^{(N)}(k) = \frac{1}{N} \sum_{r=1}^{N} x_r^{(N)}(k),$$

we get

$$x^{(N)}(k+1) = x^{(N)}(k) + \kappa \left(w - \frac{1}{N} \sum_{r=1}^{N} x_r^{(N)}(k) M_r^{(N)}(k) \right).$$

For large N, we would expect a law-of-large-numbers result of the form

$$\frac{1}{N} \lim_{N \to \infty} \sum_{r=1}^{N} x_r^{(N)}(k) M_r^{(N)}(k) = \lim_{N \to \infty} E\left(x_r^{(N)}(k) M_r^{(N)}(k) \right), \quad \text{in probability}$$

$$= E\left(x_r^2(k)(1 - e^{-\gamma x(k)}) \right),$$

where

$$x_r(k+1) = x_r(k) + \kappa \left(w - x_r(k) M_r(k) \right), \tag{8.27}$$

$$x(k+1) = x(k) + \kappa \left(w - E(x_r^2(k))(1 - e^{-\gamma x}), \right) \tag{8.28}$$

$$M_r(k) = Poi\left(x_r(k) \left(1 - e^{-\gamma x(k)} \right) \right). \tag{8.29}$$

A similar limit was established for a slightly different model in [91]. In passing, we note that $x(k)$ is deterministic, and is in fact equal to $E(x_r(k))$. However,

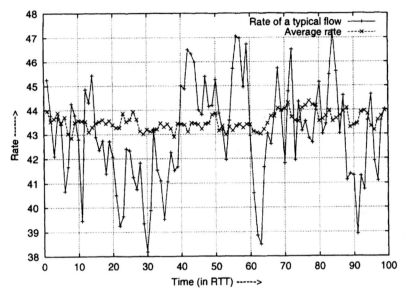

Fig. 8.4. Plot of rate as a function of time with price feedback. The number of source is 100

$x_r(k)$ is a random variable whose second moment explicitly affects the dynamics of $x(k)$.

While many authors have empirically observed that the deterministic model

$$x(k+1) = x(k) + \kappa \left(w - x^2(k)(1 - e^{-\gamma x(k)}) \right) \qquad (8.30)$$

seems to describe the average behavior of TCP-type sources well, the correct limiting result under our scaling seems to be (8.27)-(8.29). A satisfactory resolution to this discrepancy is lacking at this time. A similar discrepancy also arises in the case of queue-based marking as well. This discrepancy does not arise if the arrival process at the link is the sole source of randomness. In other words, if probabilistic marking is not used, but the exact link shadow price is conveyed to the source, then the limiting model is indeed given by (8.30). We refer the interested reader to [18], where this result was shown under a model for the packet arrival process at the link that is different from the one used in this section.

8.6 Appendix: The weak law of large numbers

The main result presented in Section 8.1 is essentially a weak law of large numbers result. We will briefly review the proof of the weak law of large numbers of independent random variables in this appendix in the hope that this would further clarify the basic idea behind the proof in Section 8.1.

Consider a set of independent, identically distributed (i.i.d.) random variables $\{X_i\}$. Let μ and σ^2 denote respectively the mean and variance of each of these random variables. The weak law of large numbers is the following theorem.

Theorem 8.5. *Let*

$$S_n = \frac{1}{n} \sum_{i=1}^{n} X_i.$$

Then $S_n \to \mu$ in probability as $n \to \infty$, i.e.,

$$\lim_{n \to \infty} P\left(|S_n - \mu| \geq \epsilon\right) = 0, \qquad \forall \epsilon > 0.$$

□

Before, we prove this weak law of large numbers, let us first establish the following lemmas.

Lemma 8.6. *(Markov's inequality) Let Y be a positive random variable. Then,*

$$P(Y \geq y) \leq \frac{E(Y)}{y}.$$

Proof. Note that

$$\begin{aligned}
E(Y) &= E(Y I_{Y \geq y}) + E(Y I_{Y < y}) \\
&\geq E(Y I_{Y \geq y}) \\
&\geq y E(I_{Y \geq y}) \\
&= y P(Y \geq y).
\end{aligned}$$

□

Lemma 8.7. *(Chebyshev's inequality) Let Z be a random variable with mean m and variance s^2. Then,*

$$P(|Z - m| \geq \varepsilon) \leq \frac{s^2}{\varepsilon}.$$

Proof. Let

$$Y := |Z - m|^2.$$

Then,

$$P(|Z - m| \geq \varepsilon) = P(Y \geq \varepsilon^2).$$

Now apply Markov's inequality to get the result.   □

Now, we are ready to prove the weak law of large numbers. First, note that $E(S_n) = \mu$ and $Var(S_n) = \sigma^2/n$ due to the i.i.d. assumption on $\{X_i\}$. Thus, applying Chebyshev's inequality to S_n gives

$$P(S_n \geq \varepsilon) \leq \frac{\sigma^2}{n}$$
$$\to 0$$

as $n \to \infty$, thus proving the desired result.

9

Connection-level Models

In this chapter, we study simple models of the dynamics of a network at the connection level, i.e., at the level of file arrivals and departures. In all the previous chapters, we assumed that the number of controlled flows in the network is a constant and studied the congestion control and resource allocation problem for a fixed number of flows. In reality, connections or files or flows arrive and depart from the network. We assume that the amount of time that it takes for the congestion control algorithms to drive the source rates close to their equilibrium is much smaller than the inter-arrival and inter-departure times of files. In fact, we assume that compared to the time-scale of connection dynamics, the congestion control algorithm operates instantaneously, providing a resource allocation dictated by the utility functions of the users of the network. In the following sections of this chapter, we will study the impact of flow-level resource allocation on connection-level stochastic stability for this time-scale separated model.

9.1 Stability of weighted proportionally-fair controllers

Consider a network in which files are arriving to route r at rate λ_r files-per-second according to a Poisson process. The file sizes are independent and exponentially distributed, with the mean file size being $1/\mu_r$ on route r. Let $n_r(t)$ be the number of files on route r at time t. It is assumed that each file on route r is allocated $\hat{x}_r(t)$ bits-per-second at time t, where the rates $\{\hat{x}_r\}$ are chosen as the optimal solution to the following weighted proportionally-fair resource allocation problem:

$$\max_{\{x_r\}} \sum_r w_r n_r \log x_r \qquad (9.1)$$

subject to

$$\sum_{r:l\in r} n_r x_r \le c_l, \qquad \forall l,$$

$$x_l \ge 0, \qquad \forall l.$$

In this optimization problem, it is assumed that all files using the same route have the same utility function $w_r \log x_r$. This connection-level assumes that, compared to the inter-arrival and departure times of files, congestion control operates at very fast time scales so that, at the connection time-scale, it appears as though the congestion controller converges instantaneously to achieve weighted proportionally-fair allocation at all times. Thus, a time-scale separation is assumed between congestion control (the fast time-scale) and file arrivals and departures (the slow time-scale).

Consider the load on any link l due to the above arrival process. If a route r uses link l, then on average, it requires λ_r/μ_r bits-per-second from link l. Thus, it is rather obvious that the network can support the traffic load only if

$$\sum_{r:l\in r} \frac{\lambda_r}{\mu_r} < c_l, \qquad \forall l. \tag{9.2}$$

What is less obvious is that this is also the sufficient condition for stochastic stability. We will explain what we mean by stochastic stability later. In the rest of this section, we will prove that the condition (9.2) is sufficient for stochastic stability.

Let us first consider the following deterministic model for the dynamics of the number of files in the system:

$$\dot{n}_r = \lambda_r - \mu_r n_r \hat{x}_r, \tag{9.3}$$

with appropriate modifications to the right-hand side to ensure that $n_r \ge 0$. We will now show that, starting from any non-negative initial condition, the deterministic model reaches the state where $n_r = 0$ for all r in finite time. To this end, consider the candidate Lyapunov function

$$V(t) = \frac{1}{2} \sum_r k_r n_r^2(t),$$

where the k_r's are non-negative constants, to be chosen appropriately later. Note that

$$\dot{V} = \sum_r k_r n_r (\lambda_r - \mu_r n_r \hat{x}_r),$$

where we have assumed that $n_r > 0$ for all r. The case where some of the n_r's are equal to zero is left as an exercise for the reader.

We now make use of the following well-known result about concave optimization (see Appendix 2.3): suppose $\hat{\mathbf{y}}$ is the optimal solution to $\max_{\mathbf{y}} f(\mathbf{y})$ subject to $\mathbf{y} \in \mathcal{S}$, where $f(\cdot)$ is a concave function and $\mathcal{S}$ is a convex set. Then, for any $\mathbf{y} \in \mathcal{S}$, the following is true:

$$\sum_i \frac{\partial f}{\partial y_i}(\mathbf{y})(y_i - \hat{y}_i) \le 0.$$

To make use of this result, let us rewrite $\dot{V}$ as

$$\dot{V} = \sum_r k_r n_r^2 \mu_r \left(\frac{\lambda_r}{\mu_r n_r} - \hat{x}_r \right).$$

Recall that $\{\hat{x}_r\}$ is the optimal solution to (9.1). Further, if we choose $x_r = \frac{\lambda_r}{\mu_r n_r}$, then this set of rates satisfy the link capacity constraints. In fact, from condition (9.2), there exists an $\varepsilon > 0$ such that

$$(1 + \varepsilon) \sum_{r:l \in r} \frac{\lambda_r}{\mu_r} \le c_l, \qquad \forall l.$$

Thus, $x_r = (1 + \varepsilon) \frac{\lambda_r}{\mu_r n_r}$ is also a rate allocation that satisfies the link capacity constraints. Thus, from the property of concave optimization discussed earlier, we have

$$\sum_r \frac{n_r w_r}{\frac{\lambda_r(1+\varepsilon)}{\mu_r n_r}} \left(\frac{\lambda_r(1+\varepsilon)}{\mu_r n_r} - \hat{x}_r \right) \le 0. \qquad (9.4)$$

Choosing $k_r = w_r / \lambda_r$, the previous inequality yields

$$\dot{V} \le -\varepsilon \sum_r w_r n_r.$$

Note that

$$\left(\sum_r w_r n_r \right)^2 \ge \sum_r w_r^2 n_r^2 = \frac{1}{2} \sum_r \frac{2 w_r^2}{k_r} k_r n_r^2 \ge \left(\min_r \frac{2 w_r^2}{k_r} \right) V(t).$$

Therefore, defining

$$\delta = \varepsilon \sqrt{\min_r \frac{2 w_r^2}{k_r}},$$

we have

$$\dot{V} \le -\delta \sqrt{V}.$$

Thus,

$$\sqrt{V(t)} - \sqrt{V(0)} \le -\frac{1}{2} \delta t.$$

Thus, $V(t)$ becomes zero in finite time, which automatically means that $\mathbf{n}(t)$ becomes $\mathbf{0}$ in finite time [12]. A similar stability result was established earlier for the case of max-min fair flows in [29] and stability was established for a specific network topology with proportionally-fair controllers in [72].

To prove stochastic stability, we first define $\zeta_r = \max_{l:l \in r} c_l$. Now, using uniformization, we consider the system only at certain event times, where the events correspond to arrival, departure and fictitious departures. Suppose that there are n_r files using route r immediately after an event. Then, with probability $\lambda_r / \sum_{s \in S}(\lambda_s + \mu_s \zeta_s)$, the next event is an arrival to route r; with probability $\mu_r n_r x_r / \sum_{s \in S}(\lambda_s + \mu_s \zeta_s)$, the next event is a departure from route r; and with probability $(\mu_r \zeta_r - \mu_r x_r n_r) / \sum_{s \in S}(\lambda_s + \mu_s \zeta_s)$, the next event is a fictitious departure from route r.

Let $V(k)$ denote the value of the Lyapunov function immediately after the kth event. For notational convenience, and without loss of generality, we assume that

$$\sum_{s \in S}(\lambda_s + \mu_s \zeta_s) = 1.$$

We note that

$$E(V(k+1) - V(k)|\mathbf{n}(k)) = \frac{1}{2}\sum_r \lambda_r k_r \left((n_r(k) + 1)^2 - n_r^2(k)\right)$$

$$+ \frac{1}{2}\sum_r I_{n_r > 0}\mu_r n_r(k)x_r k_r \left((n_r(k) - 1)^2 - n_r^2(k)\right)$$

$$= \sum_r \lambda_r k_r (n_r + \frac{1}{2})$$

$$+ \sum_r \mu_r n_r x_r I_{n_r > 0} k_r (-n_r + \frac{1}{2})$$

$$= \sum_r k_r n_r (\lambda_r - \mu_r n_r x_r) + C,$$

where the last step above is obtained by noting that $n_r x_r \leq \zeta_r$ and defining

$$C = \sum_r k_r (\lambda_r + \mu_r \zeta_r)/2 = k_r/2.$$

Using (9.4) as before, we obtain

$$E(V(k+1) - V(k)|\mathbf{n}(k)) \leq -\varepsilon \sum_r w_r n_r(k) + C,$$

which implies that

$$\varepsilon \sum_r w_r E(n_r(k)) \leq E(V(k) - V(k+1)) + C.$$

Summing both sides of the above inequality over k, we obtain

$$\varepsilon \sum_{k=1}^N \sum_r w_r E(n_r(k)) \leq E(V(1)) - E(V(N+1)) + NC$$

$$\leq E(V(1)) + NC.$$

Thus, the network is *stable-in-the-mean*, i.e.,

$$\limsup_{N \to \infty} \frac{1}{N} \sum_{k=1}^{N} \sum_{r} w_r E(n_r(k)) \leq \frac{C}{\varepsilon}.$$

One can show that this implies that the Markov chain is positive recurrent, which further implies that it has a well-defined steady-state distribution [57].

In the real Internet, while it may be reasonable for connection arrivals to be modelled as a Poisson process, it is not quite realistic to model the holding times as being exponential. However, the proof of stability for general holding time distributions is an open problem at the time of writing of this book.

9.2 Priority resource allocation

In the previous section, we showed that the system of controllers which solve a resource allocation problem with concave utility functions (at each time-instant in the connection time-scale) provide the maximum connection-level throughput. However, this may not be true in general for other resource allocation schemes. In this section, we consider a priority resource allocation scheme and show that its stable throughput region is smaller than the throughput region for the resource allocation scheme in the previous section [12].

Consider two links, say A and B of unit capacity each. The route of sources of type 0 include both links, whereas sources of type 1 use only link A and sources of type 2 use only link B. Suppose that sources of type 1 and 2 have priority over sources of type 0. In other words, at link A, a source of type 0 is served only if there are no sources of type 1 waiting for service. Similarly at link B, a source of type 0 is served only if there are no sources of type 2 waiting for service. In this case, the stability condition is given by

$$\rho_1 < 1, \qquad \rho_2 < 1, \qquad \rho_0 < (1 - \rho_1)(1 - \rho_2).$$

To see this, note that the fraction of time that link A is not serving users of type 1 is $(1 - \rho_1)$. Similarly, the fraction of time that link A is not serving users of type 2 is $(1 - \rho_2)$. Thus, the fraction of time that both links are free to serve users of type 0 is $(1 - \rho_1)(1 - \rho_2)$. Finally, the offered load of users of type 0 should be less than the amount of time that both links are free to serve it, thus giving us the result.

Under a proportionally-fair allocation of the link capacities, the stability condition is

$$\rho_0 + \rho_1 < 1, \qquad \rho_0 + \rho_2 < 1.$$

In other words,

$$\rho_0 < \min\{1 - \rho_1, 1 - \rho_2\}.$$

Since $(1 - \rho_1) < 1$ and $(1 - \rho_2) < 1$, clearly

$$(1 - \rho_1)(1 - \rho_2) < \min\{1 - \rho_1, 1 - \rho_2\}.$$

Thus, proportionally-fair allocation has a larger stability region than priority service.

10

Real-time Sources and Distributed Admission Control

In the previous chapters, we have considered only elastic users, i.e., those users that are characterized by concave utility functions. In this chapter, we will consider a specific type of inelastic users: those which require an average bandwidth (say, 1 unit) and a QoS guarantee (e.g., an upper bound on the probability of packet loss or the probability that the delay exceeds some threshold) from the network. In other words, these users would not be willing to use the network if the amount of bandwidth available is less than one unit, and further, will not use any excess bandwidth if more than one unit is provided to them. An example of such sources are voice calls, which generate packets at a rate of 8 kbps to 64 kbps, depending on the codec (coding and decoding algorithm) used, and require the delay, jitter and packet loss in the network to be small to ensure good quality of the audio at the receiver.

Since elastic users derive some utility at different data rates, it makes sense for them to adapt their transmission rates depending upon the level of congestion in the network. On the other hand, for the kind of inelastic users described above, it makes more sense to regulate their admission into the network, i.e., admit a source into the network only if sufficient bandwidth is available. There are two ways to implement such an admission control mechanism:

- *Resource reservation:* Under this method, each link in the network keeps track of its available bandwidth. When a source requests admission into the network, it sends a message to the network requesting a certain amount of bandwidth. The message traverses all the links on the source's route and if there is sufficient capacity available in all the links, then the network admits the source. If not, the source's request is denied and the source is not admitted into the network. Such a scheme requires a *signalling protocol* for the network and the sources to communicate prior to admitting each source. Further, it requires each source to keep track of the available bandwidth at all times. Thus, it requires a certain overhead which is not implemented in most Internet routers today. Therefore, in this chapter, we

will study an alternate scheme, whereby each source probes the network
for the bandwidth it needs.

• *Probing:* Upon arrival, each source sends a certain number, say m, of probe
packets into the network. The network marks or drops these packets according to some active management strategy (such as drop-tail, RED,
REM, etc.). If more than r of these packets are marked or dropped, then
the source decides not to join the network; otherwise, the source joins the
network. Thus, the admission decision is taken by the sources in a distributed manner. There is one disadvantage with this admission control
scheme, as compared to resource reservation: there is a non-zero probability of making an error in the admission process. The error could be of one
of two types: a source could be admitted into the network when there is in-
sufficient bandwidth available, or a source could be denied admission even
though there is sufficient bandwidth in the network. Under the first type
of error, the admitted source may suffer from a poor quality-of-service,
i.e., be subject to either large delays or high levels of packet loss. Under
the second type of error, the network resource utilization will suffer. In the
rest of this chapter, we will present a simplified version of the analysis of
the probing and distributed admission control scheme from [51] and show
that it is possible to maintain a desired QoS as well as to achieve high
levels of utilization.

10.1 Resource sharing between elastic and inelastic users

Consider a simple model, where inelastic users on route r arrive at rate λ_r
users/sec and a user on route r has F_r units of data to transmit. As mentioned
earlier, we assume that the user generates data at an average rate of one unit-
per-sec. Suppose the probing mechanism is used for admission control with
$r = 1$, i.e., a user is admitted if none of its m probe packets are marked. Then,
a deterministic approximation to the number of users, n_r, on route r is given
by the differential equation

$$\dot{n}_r = \lambda_r(1 - q_r)^m - \frac{n_r}{F_r}. \tag{10.1}$$

Since the data rate of inelastic users has been assumed to be 1, n_r is also
the rate at which inelastic users on route r are generating data. If p_l is the
probability of a packet being marked on link l, then

$$1 - q_r = \prod_{l \in r}(1 - p_l),$$

where $p_l = f_l(y_l)$ is a function of the arrival rate y_l at link l. Let us suppose
that an elastic user on route r has a utility function $U_r(x_r)$, and that it adjusts
its rate according to the following primal controller introduced in Chapter 3:

$$\dot{x}_r = \kappa_r(x_r)\left((1 - q_r)U_r'(x_r) - q_r\right). \tag{10.2}$$

The following theorem establishes the stability of the system of equations (10.1)-(10.2), and also shows how the network resources (i.e., link capacities) are allocated among the elastic and inelastic users.

Theorem 10.1. *The system of equations (10.1)-(10.2) is stable and converges to the unique solution of the following problem:*

$$\max_{\mathbf{x}_r, \mathbf{n}_r \geq 0} \mathcal{U}(\mathbf{x}_r, \mathbf{n}_r), \tag{10.3}$$

where

$$\mathcal{U}(\mathbf{x}_r, \mathbf{n}_r) = -\sum_r \frac{1}{m} \int_0^{n_r} \log \frac{\eta}{F_r \lambda_r} d\eta + \sum_r \int_0^{x_r} \log(1 + U_r'(\sigma)) d\sigma$$

$$+ \sum_l \int_0^{y_l} \log(1 - f_l(y)) dy, \tag{10.4}$$

where $y_l = \sum_{r:l \in r}(x_r + n_r)$.

Proof. Since $-\log(\cdot)$, $\log(1 + U_r'(\cdot))$ and $(1 - f_l(\cdot))$ are decreasing functions, it follows that $\mathcal{U}$ is a concave function. Further since $-\log(\cdot)$ and $\log(1 + U_r'(\cdot))$ are strictly decreasing, it follows that $\mathcal{U}$ is a strictly concave function. Thus, there is a unique solution to $\max_{\mathbf{x}_r, \mathbf{n}_r \geq 0} \mathcal{U}(\mathbf{x}_r, \mathbf{n}_r)$. Now,

$$\dot{\mathcal{U}} = \sum_r \left(-\frac{1}{m} \log \frac{n_r}{F_r \lambda_r} + \sum_{l:l \in r} \log(1 - f_l(y_l)) \right) \dot{n}_r$$

$$+ \sum_r \left(-\log \frac{1}{1 + U_r'(x_r)} + \sum_{l:l \in r} \log(1 - f_l(y_l)) \right) \dot{x}_r$$

$$= \sum_r \frac{\lambda_r}{m^2} \left(-\log \frac{n_r}{F_r \lambda_r} + m \log(1 - q_r) \right) \left(-\frac{n_r}{F_r \lambda_r} + (1 - q_r)^m \right)$$

$$+ \sum_r \kappa_r(x_r)(1 + U_r'(x_r))$$

$$\times \left(-\log \frac{1}{1 + U_r'(x_r)} + \log(1 - q_r) \right) \left(-\frac{1}{1 + U_r'(x_r)} + (1 - q_r) \right)$$

$$\geq 0,$$

where the last step follows from the fact that $\log(\cdot)$ is an increasing function. Further, $\dot{\mathcal{U}} = 0$ only at the solution to (10.3). Hence, $\mathcal{U}$ serves as a Lyapunov function for (10.2)-(10.1), thus establishing the desired result. $\quad\square$

10.2 Probing and distributed admission control

To understand the performance of distributed admission control schemes, we consider the following simple model: a single link accessed by inelastic users only. Further, we suppose that all the stochastic descriptions of the packet generation processes of the users are identical and independent. Our goal in this section is to compare distributed admission control with the centralized resource reservation scheme. Since a signaling protocol is used for resource reservation, the network can reserve exactly the amount needed for a user, and thus, can achieve high utilization with low packet loss and delay. Thus, it makes sense to evaluate the performance of the distributed control scheme by comparing the fraction of users that are not admitted into the network, and the probability of QoS violation, to that of the resource reservation scheme.

Let us suppose that users arrive according to a Poisson process of rate λ to a link of capacity C packets-per-second. Let us also suppose that the users specify a quality-of-service (QoS) requirement in terms of the maximum delay in the buffer at the link, and further, suppose that the network agrees to provide this QoS for at least a fraction $(1 - \varepsilon)$ of the time. Equivalently, the network guarantees that the probability of QoS violation will be less than or equal to ε. Assume that the amount of time spent by each user in the network is exponentially distributed with mean $1/\mu$, and that these durations are independent among the users. We assume the following time-scale separation between the queueing process at the link and the arrival and departure processes of the users: the queueing process reaches equilibrium between successive user arrival or departure events. Thus, if there are n users in the system, then using the statistics of the packet arrival process, one can calculate the equilibrium distribution of the queue lengths (or workload as measured by the number of bits in the buffer) and use this distribution to compute the probability that the delay exceeds the user-specified threshold. We will present models for computing this probability later in this chapter. For now, when there are n users in the network, we assume that the probability of QoS violation is given by a function $P(n)$.

Under a resource-reservation scheme, the network can compute the maximum number of users, N_{max}, that can be admitted into the network:

$$N_{max} = \max\{n : P(n) \leq \varepsilon\}.$$

The network can then admit a new source only if the number of sources already in the network is less than or equal to N_{max}. Thus, the QoS guarantee is never violated and the only quantity of interest is the utilization of the link or, equivalently, the fraction of users that are denied admission into the network. Thus, the link can be thought of as an $M/M/N_{max}/N_{max}$ loss system and the fraction of rejected calls by the resource reservation scheme, denoted by p_b, is given by the Erlang-B formula [10]:

$$p_b = \frac{\dfrac{\rho^{N_{max}}}{N_{max}!}}{\displaystyle\sum_{n=0}^{N_{max}} \dfrac{\rho^n}{n!}}. \qquad (10.5)$$

To study the distributed admission control scheme, we assume that each probe packet is marked to indicate congestion with probability $p_c(n)$ when there are n users in the network. During the probing process, we make the assumption that each packet is marked independently with this probability. Recall that a user sends m probe packets and joins the network if less than or equal to t packets are marked. Thus, the probability that a user is admitted into the system when there are n calls already in progress, denoted by $a(n)$, is given by

$$a(n) = \sum_{i=0}^{t} \binom{m}{i} p_c^i(n)(1 - p_c(n))^{m-i}.$$

Thus, the number of users in the network $\{n(t)\}$ under the distributed admission control scheme is a Markov chain where the transition from state n to state $n+1$ occurs at rate $\lambda a(n)$ and the transition from state n to $n-1$ occurs at rate $n\mu$. In the marking models to be presented later in this chapter, the call admission probability $a(n)$ becomes zero for n larger than some value N_{max}. Thus, the Markov chain describing the evolution of the number of calls in progress under the distributed admission control scheme has a finite number of states, taking integer values from 0 to N_{dist}. This Markov chain is shown in Figure 10.1.

Fig. 10.1. Markov chain describing the evolution of the distributed admission control scheme

Let $\pi(n)$ denote the steady-state probability that there are n calls in the system under the distributed admission control scheme. Then, $\pi(n)$ is given by the following detailed balanced equations [10, 47]:

$$\lambda a(n)\pi(n) = (n+1)\mu\pi(n+1), \qquad (10.6)$$

along with the condition

$$\sum_{n=1}^{N_{dist}} \pi(n) = 1.$$

Since $\pi(n)$ is the probability that there are n users in the network and $(1 - a(n))$ is the probability that an arriving user is blocked in state n, the fraction of users blocked under the distributed admission control scheme, denoted by p_{dac}, is given by

$$p_{dac} = \sum_{n=0}^{N_{dist}-1} \pi(n)(1 - a(n)). \qquad (10.7)$$

Unlike the centralized admission control scheme, it is possible for the number of users admitted into the network to be larger than the number required to guarantee the QoS. In other words, if $N_{dist} > N_{max}$, then with a non-zero probability, the number of users in the system is too high. We quantify the violation by the fraction of time that the number of users in the system is greater than N_{max}. This quantity, denoted by p_{qos}, is given by

$$p_{qos} = \sum_{n=N_{max}+1}^{N_{dist}} \pi(n). \qquad (10.8)$$

Clearly, the challenge in the distributed admission control is to design the congestion indication (marking) scheme at the link to keep p_{qos} small while maintaining the blocking probability p_{dac} close to the centralized blocking probability p_b. In fact, we will see that, when the link capacity is large, this is not very hard to achieve. We will also see that the parameters of the distributed admission control scheme (m, r) and the parameters of the congestion indication mechanism at the link are insensitive to the load (λ/μ) and thus, one can design the system parameters easily to approach the performance of centralized admission control scheme using the probing method.

Note that the Markov chain in Figure 10.1 and the loss system in the resource-reservation scheme are both reversible Markov chains, and hence the steady-state distributions of the number of users in the network under both schemes are independent of the distribution of the amount of time spent by each user in the network [47]. Thus, our assumption that the service-time distribution of the users is exponentially distributed is not required; however, we still require the arrival process to be Poisson for the expressions for $\{\pi(n)\}$ to be valid. Next, we present models for marking and buffer flow that will allow us to compute the quantities $P(n)$ and $p_c(n)$.

10.3 A simple model for queueing at the link buffer

Suppose that there are n calls in progress in the network. To describe the queueing process at the link buffer, we have to assume something about the packet arrival process at the link. Let us suppose that each user generates packets at the rate of λ_p packets-per-second. Then, the number of packets in a time interval of length τ is given by $n\lambda_p\tau$. Further, let the variance of the number of packets in a time interval of length τ be given by $n\sigma^2\tau$.

Now, we make the assumption that the number of arriving packets in a time-interval τ has a Gaussian distribution and the number of arriving packets in disjoint time intervals are independent. Such an approximation is called the *diffusion approximation* and the reader is referred to the appendix for the motivation behind such an approximation. Under this approximation, the number of packets in the buffer is assumed to be a real number, rather than an integer, and the probability that the number of packets is larger than x is given by

$$F_n(x) = e^{\frac{-2(C - n\lambda_p)x}{n\sigma^2}}, \tag{10.9}$$

where we have assumed that $n\lambda_p < C$. If $n\lambda_p \geq C$, then $F_n(x) \equiv 1$. If the QoS requirement for the users is that the delay in the link buffer should not exceed D, then it is equivalent to stating that the number of packets in the buffer should not exceed DC. Thus, $P(n)$ is given by

$$P(n) = F_n(DC).$$

To compute $p_c(n)$, we assume that a packet is marked with probability $f_c(x)$ when there are x packets in the buffer. If, for example, REM is used to signal congestion, then

$$f_c(x) = 1 - e^{-\theta x},$$

for some $\theta > 0$. Further, marking could be done based on the number of packets in the real queue or the number of packets in a virtual queue, where the capacity of the virtual queue is $\tilde{C} = \gamma C$, for some $\gamma \leq 1$. If γ is taken to be 1, then the marking is based on the number of packets in the real queue, whereas if $\gamma < 1$, then the marking is based on the number of packets in the virtual queue. Thus, to get the complementary cumulative distribution of the number of packets in the virtual buffer, we just simply replace C in (10.9) by γC to obtain

$$\tilde{F}_n(x) = e^{\frac{-2(\gamma C - n\lambda_p)x}{n\sigma^2}}.$$

Thus, the probability density function of the number of packets in the virtual buffer is given by

$$\tilde{f}_n(x) = \frac{-2(\gamma C - n\lambda_p)}{\sigma^2} e^{\frac{-2(\gamma C - n\lambda_p)x}{n\sigma^2}}.$$

Thus, the marking probability $p_c(n)$ is given by

$$p_c(n) = \int_0^\infty f_c(x)\tilde{f}_n(x)dx.$$

For example, if REM is used to mark packets, then the marking probability when there are n users in the network is given by

$$p_c(n) = \frac{n\sigma^2\theta}{n\sigma^2\theta + 2(\gamma C - n\lambda_p)}.$$

Numerical results in [51] show that the difference between using probing and centralized control, using the models presented above, is negligible for large systems. In the above models, we have ignored the effect of probing calls. In other words, the queue dynamics at the link is not only influenced by the number of users already in the network, but also the number of users that are probing the system for available bandwidth. This effect can be noticeable especially when the load on the network is high. In such cases, the choice of the marking process at the link could make a significant difference to the performance of the network; see [15, 30] for details.

10.4 Appendix: Diffusion approximation

The central limit theorem (CLT) for random variables provides an approximation to the behavior of sums of random variables using only the first two moments. Similar approximations for random processes which provide sparse descriptions using the mean and auto-covariance functions are called *functional central limit theorems*, i.e., central limit theorems for random functions.

Consider a discrete-time arrival process $\{a(k)\}$, where $a(k)$ could, for example, represent the number of arriving bits at time k. Let

$$A(k) := \sum_{i=1}^{k} a(k)$$

represent the total number of arriving bits up to time k. Let us suppose that the process is wide-sense stationary with $\mu := E(a(k))$ and

$$C(k) := Cov(a(n), a(n+k)) = E((a(n) - \mu)(a(n+k) - \mu)).$$

Analogous to the central limit theorem for random variables, a natural approximation to seek for $A(k)$ would be a Gaussian process. A Gaussian process with a simple description is the *Brownian motion*, also known as the *Wiener process*, which is a continuous-time process defined as follows:

Definition 10.2. *A Brownian motion $\{W(t), \quad t \geq 0\}$ with drift μ and infinitesimal variance σ^2 is a sample-path continuous process which has the following properties:*

- *it is an independent-increment process, i.e., $W(t_2) - W(t_1)$ and $W(t_4) - W(t_3)$ are independent whenever $[t_1, t_2]$ and $[t_3, t_4]$ are disjoint intervals, i.e., increments over disjoint intervals are independent.*
- *$W(0) = 0$, and $W(t)$ has a Gaussian distribution with mean μt and variance $\sigma^2 t$.*

A standard Brownian motion is a process for which $\mu = 0$ and $\sigma^2 = 1$. $\square$

We are seeking to approximate a discrete-time process by a continuous-time process. Thus, the first "trick" is to convert the discrete-time process into a continuous-time process by defining a new process $\{A^{(n)}(t)\}$ as follows:

$$A_n(t) := A(\lfloor nt \rfloor),$$

where n is a scaling parameter that compresses time so that the discrete time steps appear continuous as n becomes large. To understand this scaling graphically, suppose that we plot $A(k)$ as a function of k. Then, as n is increased $A_n(t)$ is a graph of the same function viewed from afar. The second "trick" is to apply the CLT scaling to try to obtain a Gaussian approximation. To do this, let us define

$$\sigma_n^2(t) = \frac{1}{\lfloor nt \rfloor} \text{Var}(A_n(t)).$$

Then define $\{S_n(t)\}$ to be

$$S_n(t) = \frac{A_n(t) - \mu \lfloor nt \rfloor}{\sqrt{n \sigma_n^2(t)}}.$$

Often, first courses in probability deal with the CLT for independent random variables. However, under some mild additional assumptions beyond the assumption that $\lim_{n \to \infty} \sigma_n^2(t)$ exists, the CLT is also applicable to weakly dependent random variables [92].

We will assume that $\lim_{n \to \infty} \sigma_n^2(t) = \sigma^2$ where $0 < \sigma^2 < \infty$. Under this assumption, $\lim_{n \to \infty} Var(S_n(t)) = t$. Thus, it is reasonable to expect that $S_n(t)$ converges in an appropriate sense to a Brownian motion. Indeed this is true under some assumptions, which we will not explore here [97]. We will simply assume that the covariance function of our process satisfies the conditions under which convergence occurs. Note that, to characterize the limiting Brownian motion, we only need μ and σ^2.

10.4.1 Brownian motion through a queue

Let $a(k)$ be the number of bits arriving to a queue at time k. Let $c(k)$ be the capacity of the link draining the queue at time k, i.e., the link can serve a maximum of $c(k)$ bits at time k. The processes $\{a(k)\}$ and $\{c(k)\}$ are assumed to be stationary. The number of bits in the queue at time k evolves according to the equation

$$q(k + 1) = (q(k) - c(k))^+ + a(k).$$

Let us suppose that $q(0) = 0$ and that we are interested in the steady-state probability that the queue length (in bits) exceeds a certain value B. In other words, we wish to compute $\lim_{k \to \infty} P(q(k) > B)$. For the steady-state distribution to exist, we assume the stability condition, $E(a(1)) < E(c(1))$. We first note that

$$q(1) = \max\{a(0) - c(0), 0\},$$
$$q(2) = \max\{q(1) + a(1) - c(1), 0\}.$$

Thus,

$$q(2) = \max\{a(0) - c(0) + a(1) - c(1), a(1) - c(1), 0\}.$$

Defining

$$A(k) = \sum_{i=0}^{k-1} a(i), \quad \text{and} \quad C(k) = \sum_{i=0}^{k} c(i),$$

we can rewrite $a(2)$ as

$$q(2) = \max\{A(2) - C(2), A(2) - C(2) - (A(1) - C(1)), 0\}$$
$$= A(2) - C(2) - \min_{0 \le i \le 1} [A(i) - C(i)],$$

where $A(0)$ and $C(0)$ are defined to be zero. Continuing as above, we get the equation

$$q(k) = V(k) - \min_{0 \le i \le k} V(i), \tag{10.10}$$

where $V(k) := A(k) - C(k)$. See Figure 10.2 for a pictorial view of the relationship between $V(k)$ and $q(k)$. To use the Brownian motion approximation of the previous subsection, we let the system operate in heavy traffic as follows: consider a sequence of queueing systems, indexed by a parameter n, such that

$$\lambda_n - \mu_n = -\frac{m}{\sqrt{n}},$$

where $m > 0$ is some constant. As $n \to \infty$, it is clear that $\lambda_n \to \mu_n$.

It is not difficult to see that, in the limit $n \to \infty$, the mean and variance of

$$V_n(t) := \frac{V(\lfloor nt \rfloor)}{\sqrt{n}}$$

are given by $-m$ and σ^2, where

$$\sigma^2 = \lim_{k \to \infty} \frac{\text{Var}(V(k))}{k}.$$

Further, $V_n(t)$ converges to a Gaussian distribution. Thus, it is reasonable to expect that $V_n(t)$ converges to a Brownian motion with drift m and infinitesimal variance σ^2. In fact, we will assume that this convergence indeed occurs in an appropriate sense. Assuming this, we will now proceed to compute the probability $P(q_n(t)/\sqrt{n} > x)$, where $q_n(t) = q(\lfloor nt \rfloor)$. Note that

$$P\left(\frac{q_n(t)}{\sqrt{n}} > x\right) = P\left(V_n(t) - \min_{s = 0, 1/n, 2/n, \ldots, \frac{\lfloor nt \rfloor}{n}} \frac{V_n(s)}{\sqrt{n}} > x\right).$$

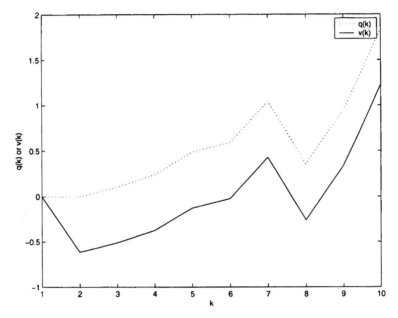

Fig. 10.2. The relationship between $V(k)$ and $q(k)$. Note that $\min_{0 \leq k \leq 10} V(k) = V(2)$. Thus, $q(10) = V(10) - V(2)$.

Therefore, assuming $X(t) = \lim_{n \to \infty} q_n(t)/\sqrt{n}$ exists in an appropriate sense, the above equation suggests the approximation

$$P(X(t) > x) = P(W(t) - \inf_{0 \leq s \leq t} W(s) > x),$$

where $W(t)$ is a Brownian motion with parameters $-m$ and σ^2.

The mapping from a function $\alpha(t)$ to another function $\beta(t)$ given by

$$\beta(t) = \alpha(t) - \inf_{0 \leq s \leq t} \alpha(s)$$

is called a *reflection* mapping. Hence,

$$X(t) = W(t) - \inf_{0 \leq s \leq t} W(s)$$

is called a *reflected Brownian motion* (RBM, for short).

10.4.2 A lower bound for $\lim_{t \to \infty} P(q(t) > x)$

We first note that

$$P(X(t) > x) = P(W(t) - \inf_{0 \le s \le t} W(s) > x))$$

$$= P(\sup_{s \in [0,t]} W(t) - W(s) > x)$$

$$\ge \sup_{s \in [0,t]} P(W(t) - W(s) > x)$$

$$= \sup_{s \in [0,t]} Q(\frac{x + m(t - s)}{\sigma \sqrt{t - s}})$$

$$= Q\left(\inf_{u \in [0,t]} \frac{x + mu}{\sigma \sqrt{u}} \right)$$

where $Q(\cdot)$ is the complementary cumulative distribution function of a Gaussian random variable with mean 0 and variance 1. The third line above follows from the fact that the probability of a union of events is lower bounded by the probability of any one of the events, and the fourth line follows from the definition of $W(t)$ and the last line follows from the fact that $Q(\cdot)$ is a decreasing function. Thus,

$$\lim_{t \to \infty} P(X(t) > x) \ge Q(\inf_{u \ge 0} \frac{x + mu}{\sigma \sqrt{u}}).$$

Straightforward minimization shows that the infimum is achieved at the point $u^* = x/m$, and thus,

$$\lim_{t \to \infty} P(X(t) > x) \ge Q\left(\frac{2\sqrt{xm}}{\sigma} \right).$$

We now using the following well-known approximation

$$Q(y) \approx \frac{1}{\sqrt{2\pi}(1 + y)} e^{-y^2/2} \sim e^{-y^2/2},$$

where $\sim$ denotes logarithmic equivalence. In other words, $f(x) \sim g(x)$ if $\lim_{x \to \infty} \log f(x) / \log g(x) = 1$. Thus, we arrive at the following (logarithmic) approximation for the lower bound [79]:

$$\lim_{t \to \infty} P(X(t) > x) \ge e^{-2xm/\sigma^2}.$$

If this approximation is correct, then it suggests that the tail probability decays exponentially fast in x and the rate of decay is a simple function of the first two moments. Indeed this approximation is precise as we will see in the next subsection.

10.4.3 Computing the steady-state density using PDE's

Let us compute the distribution of $q(t)$, the number of bits in a queue at time t, where the difference between the arrival process and the potential service process is a Brownian motion. In other words,

$$X(t) = W(t) - \inf_{0 \leq s \leq t} W(s),$$

where $W(t)$ is a Brownian motion with drift $-m$ and infinitesimal variance σ^2. For $s < t$, let $p_{s,t}(x_s, x_t)$ denote the pdf (probability density function) of $X(t)$ given that $X(s) = x_s$, i.e.,

$$p_{s,t}(x_s, x_t) = \frac{d}{dx_t} P(X(t) \leq x_t | X(s) = x_s).$$

We now present a heuristic derivation of the partial differential equation satisfied by $p_{s,t}(\cdot, \cdot)$ along the lines of the derivation in [88].

First, we note that $p_{s,t}(\cdot, \cdot)$ satisfies the *Chapman-Kolmogorov* equation

$$p_{s,t}(y, x) = \int_0^{\infty} p_{s,\tau}(y, w) p_{\tau,t}(w, x) dw,$$

for any $\tau \in (s, t)$. Next, due to the fact that the input process is a Brownian motion, we observe that $p_{\tau,t}$ is $\phi(-m(t - \tau), \sigma^2(t - \tau); x - w)$, where $\phi(a, b; y)$ denotes a Gaussian pdf with mean a, variance b, evaluated at the point y. Thus,

$$\begin{aligned}
p_{0,t}(y, x) &= \int_0^{\infty} p_{0,t-\delta}(y, w) p(t - \delta, t)(w, x) dw \\
&= \int_0^{\infty} \left[p_{0,t}(y, x) + \frac{\partial p_{0,t}(y, x)}{\partial x_t}(w - x) - \frac{\partial p_{0,t}(y, x)}{\partial t} \delta \right. \\
&\quad \left. + \frac{1}{2} \frac{\partial^2 p_{0,t}(y, x)}{\partial x_t^2}(w - x)^2 + \cdots \right] \phi(-m\delta, \sigma^2 \delta; x - w) dw,
\end{aligned}$$

$$(10.11)$$

where the second step above follows from a Taylor's series expansion of $p_{0,t-\delta}(y, w)$. Next, we note that, for small δ, a Gaussian random variable with mean $-m\delta$ and standard deviation $\sigma\sqrt{\delta}$ has most of its probability density concentrated in a small interval of length $O(\sqrt{\delta})$ around $-m\delta$. Thus,

$$\int_0^{\infty} \phi(-m\delta, \sigma^2 \delta; x - w) dw \approx 1,$$

$$\int_0^{\infty} (x - w) \phi(-m\delta, \sigma^2 \delta; x - w) dw \approx -m\delta,$$

$$\int_0^{\infty} (x - w)^2 \phi(-m\delta, \sigma^2 \delta; x - w) dw \approx \sigma^2 \delta.$$

Dividing (10.11) by δ and letting $\delta \to 0$ yields

$$-\frac{\partial p_{0,t}(y, x)}{dt} = m \frac{\partial p_{0,t}(y, x)}{\partial x_t} + \frac{\sigma^2}{2} \frac{\partial p_{0,t}^2(y, x)}{\partial^2 x_t^2},$$

which is known as the *forward* diffusion equation or the *Fokker-Planck* equation.

To compute the steady-state density, set $\frac{\partial p_{0,t}(y,x)}{\partial t} = 0$, since, by definition, the steady-state density does not change with time. For the same reason, we will also drop the reference to time and initial condition and simply denote the pdf as $p(x)$. Thus, we have the following second-order ordinary differential equation for the steady-state pdf:

$$-m\frac{dp}{dx} = \frac{\sigma^2}{2}\frac{d^2p}{dx^2}.$$

The solution to the above equation is given by

$$p(x) = C_1 + C_2 e^{-\frac{2mx}{\sigma^2}}$$

for some constants C_1 and C_2. Noting that $p(x)$ should go to zero as $x \to \infty$, we get $C_1 = 0$. Using the condition $\int_0^\infty p(x)dx = 1$ yields

$$p(x) = \frac{2m}{\sigma^2} e^{-\frac{2mx}{\sigma^2}}.$$

11

Conclusions

In this book, we have summarized the exciting advances in the field of mathematical modelling of Internet congestion control over the last few years. As mentioned in the Introduction, this subject is vast and we have confined our attention to only those topics that lead to simple, decentralized schemes. We have concentrated our attention on algorithms that directly result from a convex programming view of congestion control as a mechanism for resource allocation. While this viewpoint is sufficiently rich in allowing us to model many practical aspects of window-flow control as implemented in the Internet, it does not address two important algorithms implemented within TCP, namely, *slow start* and *timeout*. For practical implementation of any congestion controller, the slow-start phase and a timeout mechanism are important and are worth further study. We have also not characterized the robustness of virtual queue-based marking mechanisms in this book. Compared to marking based on the contents of the real-queue, virtual-queue-based marking has many advantages. The reader is referred to [32, 31, 51, 66] for an analysis of the properties of virtual-queue-based congestion feedback. Another important problem not discussed in this book is the development of algorithms to limit the impact of users who do not obey congestion signals on the users who reduce their rates in response to congestion feedback from the network. This problem has been addressed in [86, 85, 24].

From a design point of view, we have emphasized the use of linear control theory to design the source and link parameters. At the time of writing of this book, a global stability analysis of a heterogeneous, general topology network is lacking. This is an important open problem which may be closely related to the modelling of the slow-start mechanism. As pointed out in Jacobson's paper on his congestion control algorithm [39], one may be able to view slow start as a mechanism for moving the system close to equilibrium and the congestion-avoidance phase as a mechanism for asymptotic stability. Thus, the controller should perhaps be viewed as having two sets of dynamics: one near the equilibrium and one further away from the equilibrium, somewhat like the inverted pendulum and double-inverted pendulum models that are well known to control theorists[1] (see [23] and references within).

The connection between stochastic models and their deterministic limits has also not been fully resolved. Generalizations of the results in Chapter 8 to general topology networks, heterogeneous delays and arbitrary utility functions are all open problems. It is important to answer these problems to fully justify the design criteria obtained by considering deterministic models.

Another key open problem in our opinion is the proof of stochastic stability of the connection-level model of TCP for non-exponential holding times. As pointed out in Chapter 9, if the resource allocation scheme is not properly designed, then the resulting throughput may suffer. Thus, it is important to characterize the stability region for the convex optimization-based resource allocation which has been shown to provide the largest possible stability region for the case of exponentially distributed file sizes.

---

[1] We thank Randy Freeman for providing the inverted pendulum analogy.

References

1. T. Alpcan and T. Başar, *Global stability analysis of end-to-end congestion control schemes for general topology networks with delay*, 2003, Preprint.
2. _____, *A utility-based congestion control scheme for internet-style networks with delay*, Proceedings of IEEE Infocom (San Francisco, California), March-April 2003.
3. E. Altman, K. Avrachenkov, C. Barakat, and R. Nunez-Queija, *State-dependent M/G/1 type queueing analysis for congestion control in data networks*, Computer Networks **39** (2002), no. 6, 789–808.
4. E. Altman, T. Başar, and R. Srikant, *Congestion control as a stochastic control problem with action delays*, Automatica (1999), 1937–1950, Special Issue on Control Methods for communication networks, V. Anantharam and J. Walrand, editors.
5. S. Athuraliya, V. H. Li, S. H. Low, and Q. Yin, *REM: Active queue management*, IEEE Network **15** (2001), 48–53.
6. F. Baccelli and D. Hong, *Interaction of TCP flows as billiards balls*, 2002, INRIA Rocquencourt, Technical Report.
7. D. Bansal and H. Balakrishnan, *Binomial congestion control algorithms*, Proceedings of IEEE Infocom, April 2001, pp. 631–640.
8. L. Benmohamed and S. M. Meerkov, *Feedback control of congestion in packet switching networks: The case of a single congested node*, IEEE/ACM Transactions on Networking **1** (1993), no. 6, 693–707.
9. D. Bertsekas, *Nonlinear programming*, Athena Scientific, Belmont, MA, 1995.
10. D. Bertsekas and R. Gallager, *Data networks*, Prentice Hall, Englewood Cliffs, NJ, 1987.
11. F. G. Boese, *Some stability charts and stability conditions for a class of difference-differential equations*, Zeitschrift für Angewandte Mathematik und Mechanik **67** (1987), 56–59.
12. T. Bonald and L. Massoulie, *Impact of fairness on Internet performance*, Proceedings of ACM Sigmetrics, 2001.
13. F. Bonomi, D. Mitra, and J. B. Seery, *Adaptive algorithms for feedback-based flow control in high speed, wide area ATM networks*, IEEE Journal on Selected Areas in Communications (1995), 1267–1283.
14. L.S. Bramko and L.L. Peterson, *TCP Vegas: end-to-end congestion avoidance on a global Internet*, IEEE Journal on Selected Areas in Communications (1995), 1465–1480.

15. L. Breslau, E. Knightly, S. Shenker, I. Stoica, and H. Zhang, *Endpoint admission control: architectural issues and performance*, Proceedings of ACM SIGCOMM 2000 (Stockholm, Sweden), August 2000.

16. D.M. Chiu and R.Jain, *Analysis of the increase and decrease algorithms for congestion avoidance in computer networks*, Computer Networks and ISDN Systems **17** (1989), 1–14.

17. B. S. Davie and L. L. Peterson, *Computer networks: A systems approach*, Morgan-Kaufman, 1999.

18. S. Deb, S. Shakkottai, and R. Srikant, *Stability and convergence of TCP-like congestion controllers in a many-flows regime*, Proceedings of Infocom, 2003.

19. S. Deb and R. Srikant, *Global stability of congestion controllers for the internet*, IEEE Transactions on Automatic Control **48** (2003), no. 6, 1055–1060.

20. ———, *Rate-based versus queue-based models of congestion control*, 2003, Technical Report.

21. C. A. Desoer and Y. T. Wang, *On the generalized Nyquist stability criterion*, IEEE Transactions on Automatic Control **25** (1980), 187–196.

22. X. Fan, M. Arcak, and J. Wen, l_p *stability and delay robustness of network flow control*, 2003, Preprint.

23. I. Fantoni, R. Lozano, and M. W. Spong, *Energy based control of the pendubot*, IEEE Transactions on Automatic Control **AC-45** (2000), no. 4, 725–729.

24. W. Feng, D. Kandlur, D. Saha, and K. Shin, *The blue queue management algorithms*, IEEE/ACM Transactions on Networking **10** (2002), no. 4, 513–528.

25. S. Floyd, *TCP and explicit congestion notification*, ACM Computer Communication Review **24** (1994), 10–23.

26. ———, *Highspeed TCP for large congestion windows*, 2002, Internet draft, draft-floyd-tcp-highspeed-01.txt.

27. S. Floyd and V. Jacobson, *Random early detection gateways for congestion avoidance*, IEEE/ACM Transactions on Networking (1993), 397–413.

28. G. F. Franklin, J. D. Powell, and A. Emami-Naeini, *Feedback control of dynamic systems*, Prentice-Hall, 2002.

29. T.-J. Lee G. de Veciana and T. Konstantopoulos, *Stability and performance analysis of networks supporting elastic services*, IEEE/ACM Transactions on Networking **9** (2001), no. 1, 2–14.

30. A. J. Ganesh, P. B. Key, D. Polis, and R. Srikant, *Congestion notification and probing mechanisms for endpoint admission control*, 2003, Coordinated Science Lab Technical Report, University of Illinois.

31. R. J. Gibbens and F. P. Kelly, *Distributed connection acceptance control for a connectionless network*, Proc. of the 16th Intl. Teletraffic Congress (Edinburgh, Scotland), June 1999, pp. 941–952.

32. ———, *Resource pricing and the evolution of congestion control*, Automatica **35** (1999), 1969–1985.

33. G. R. Grimmett and D. R. Stirzaker, *Probability and random processes*, Oxford University Press, Second Edition, 1992.

34. J. Hale and S. M. V. Lunel, *Introduction to functional differential equations*, 2nd edition, Springer Verlag, New York, NY, 1991.

35. N. D. Hayes, *Roots of the transcendental equation associated with a certain differential difference equation*, Journal of the London Mathematical Society **25** (1950), 226–232.

36. C. V. Hollot and Y. Chait, *Nonlinear stability analysis for a class of TCP/AQM schemes*, Proceedings of the IEEE Conference on Decision and Control, December 2001.

37. C.V. Hollot, V. Misra, D. Towsley, and W. Gong, *On designing improved controllers for AQM routers supporting TCP flows*, Proceedings of IEEE Infocom (Anchorage, Alaska), April 2001, pp. 1726–1734.

38. O. C. Imer, S. Compans, T. Başar, and R. Srikant, *ABR congestion control in ATM networks*, IEEE Control Systems Magazine **21** (2001), no. 1, 38–56.

39. V. Jacobson, *Congestion avoidance and control*, ACM Computer Communication Review **18** (1988), 314–329.

40. J. M. Jaffe, *A dentralized "optimal" multiple-user flow control algorithm*, IEEE Transactions on Communications (1981), 954–962.

41. C. Jin, D. Wei, S. H. Low, G. Buhrmaster, J. Bunn, D. H. Choe, R. L. A. Cottrell, J. C. Doyle, W. Feng, O. Martin, H. Newman, F. Paganini, S. Ravot, and S. Singh, *FAST TCP: From theory to experiments*, April 2003.

42. R. Johari and F. P. Kelly, *Charging and rate control for elastic traffic: Correction to published version*, Addendum to "Charging and rate control for elastic traffic," available at http://www.statslab.cam.ac.uk/ frank.

43. R. Johari and D. Tan, *End-to-end congestion control for the Internet: Delays and stability*, IEEE/ACM Transactions on Networking **9** (2001), no. 6, 818–832.

44. S. Kalyanaraman, R. Jain, S. Fahmy, R. Goyal, and B. Van dalore, *The ERICA switch algorithm for ABR traffic management in ATM networks*, IEEE/ACM Transactions on Networking (2000), 87–98.

45. A. Karbowski, *Comments on "Optimization flow control I: Basic algorithm and convergence"*, IEEE/ACM Transactions on Networking **11** (2002), no. 2, 338–339.

46. D. Katabi, M. Handley, and C. Rohrs, *Internet congestion control for future high bandwidth-delay product environments*, Proceedings of ACM SIGCOMM (Pittsburgh, PA), August 2002.

47. F. P. Kelly, *Reversibility and stochastic networks*, John Wiley, New York, NY, 1979.

48. _____, *Charging and rate control for elastic traffic*, European Transactions on Telecommunications **8** (1997), 33–37.

49. _____, *Models for a self-managed Internet*, Philosophical Transactions of the Royal Society **A358** (2000), 2335–2348.

50. _____, *Fairness and stability of end-to-end congestion control*, European Journal of Control **9** (2003), 149–165.

51. F. P. Kelly, P. B. Key, and S. Zachary, *Distributed admission control*, IEEE Journal on Selected Areas in Communications **18** (2000), 2617–2628.

52. F. P. Kelly, A. Maulloo, and D. Tan, *Rate control in communication networks: shadow prices, proportional fairness and stability*, Journal of the Operational Research Society **49** (1998), 237–252.

53. T. Kelly, *Scalable TCP: Improving performance in highspeed wide area networks*, December 2002.

54. S. Keshav, *An engineering approach to computer networks*, Addison-Wesley, Reading, MA, 1997.

55. H. Khalil, *Nonlinear systems*, 2nd edition, Prentice Hall, Upper Saddle River, NJ, 1996.

56. A. Kumar, *Comparative performance analysis of versions of TCP in a local network with a lossy link*, IEEE/ACM Transactions on Networking **6** (1998), 485–498.

57. P.R. Kumar and S.P. Meyn, *Stability of queueing networks and scheduling policies*, IEEE Transactions on Automatic Control **40** (1995), 251–260.

58. S. Kunniyur and R. Srikant, *Analysis and design of an adaptive virtual queue algorithm for active queue management*, Computer Communication Review (2001), 123–134.

59. _____, *Designing AVQ parameters for a general topology network*, Proceedings of the Asian Control Conference (Singapore), September 2002.

60. _____, *Note on the stability of the AVQ scheme*, Proceedings of the Conference on Information Sciences and Systems (Princeton, NJ), March 2002.

61. _____, *A time-scale decomposition approach to adaptive ECN marking*, IEEE Transactions on Automatic Control **47** (2002), no. 6, 882–894.

62. _____, *End-to-end congestion control: utility functions, random losses and ECN marks*, IEEE/ACM Transactions on Networking **7** (2003), no. 5, 689–702.

63. _____, *Stable, scalable, fair congestion control and AQM schemes that achieve high utilization in the internet*, IEEE Transactions on Automatic Control (2003), To appear.

64. R. La and V. Anantharam, *Utility based rate control in the internet for elastic traffic*, IEEE/ACM Transactions on Networking **10** (2002), no. 2, 272–286.

65. T. V. Lakshman and U. Madhow, *The performance of TCP/IP for networks with high bandwidth-delay products and random loss*, IEEE/ACM Transactions on Networking (1997), 336–350.

66. A. Lakshmikantha, C. Beck, and R. Srikant, *Robustness of real and virtual queue based active queue management schemes*, Proceedings of the American Control Conference, June 2003.

67. S. Liu, *Stability of high-throughput TCP with AVQ*, 2002, Technical Report, University of Illinois.

68. S. Liu, T. Başar, and R. Srikant, *Controlling the Internet: A survey and some new results*, Proceedings of IEEE Conference on Decision and Control (Maui, Hawaii), December 2003.

69. S. H. Low and D. E. Lapsley, *Optimization flow control, I: Basic algorithm and convergence*, IEEE/ACM Transactions on Networking (1999), 861–875.

70. S. H. Low, L. Peterson, and L. Wang, *Understanding vegas: A duality model*, Journal of ACM **49** (2002), no. 2, 207–235.

71. L. Massoulie, *Stability of distributed congestion control with heterogenous feedback delays*, IEEE Transactions on Automatic Control **47** (2002), 895–902.

72. L. Massoulie and J. Roberts, *Bandwidth sharing and admission control for elastic traffic*, Telecommunication Systems **15** (2000), 185–201.

73. M. Mathis, J. Semke, J. Mahdavi, and T. Ott, *The macroscopic behavior of the TCP congestion avoidance algorithm*, Computer Communication Review **27** (1997).

74. S. McCanne, V. Jacobson, and M. Vetterli, *Receiver-driven layered multicast*, Proceedings of ACM Sigcomm, 1996.

75. A. Misra, J. Baras, and T. J. Ott, *Generalized TCP congestion avoidance and its effect on bandwidth sharing and availability*, Proceedings of IEEE Globecom, 2000.

76. V. Misra, W. Gong, and D. Towsley, *A fluid-based analysis of a network of AQM routers supporting TCP flows with an application to RED*, Proceedings of ACM Sigcomm (Stockholm, Sweden), September 2000.

77. J. Mo, R. J. La, V. Anantharam, and J. Walrand, *Analysis and comparison of TCP Reno and Vegas*, Proceedings of Infocom, March 1999, pp. 1556–1563.

78. J. Mo and J. Walrand, *Fair end-to-end window-based congestion control*, IEEE/ACM Transactions on Networking **8** (2000), no. 5, 556–567.

79. I. Norros, *A storage model with self-similar input*, Queueing Systems **16** (1994), 387–396.

80. T. J. Ott, *ECN protocols and TCP paradigm*, 1999, Available at the web site http://web.njit.edu/ ott/Papers/index.html.

81. J. Padhye, V. Firoiu, D. Towsley, and J. Kurose, *Modeling TCP throughput: A simple model and its empirical validation*, Proceedings of ACM SIGCOMM, 1998.

82. F. Paganini, *A global stability result in network flow control*, Systems and Control Letters **46** (2002), no. 3, 153–163.

83. F. Paganini, J. Doyle, and S. Low, *Scalable laws for stable network congestion control*, Proceedings of the IEEE Conference on Decision and Control (Orlando, FL), December 2001, pp. 185–190.

84. F. Paganini, Z. Wang, J. Doyle, and S. Low, *A new TCP/AQM for stable operation in fast networks*, Proceedings of the IEEE Infocom (San Francisco, CA), April 2003.

85. R. Pan, C. Nair, B. Yang, and B. Prabhakar, *Packet dropping schemes, some examples and analysis*, Proceedings of the 39th Annual Allerton Conference on Communication, Control and Computing, October 2001, pp. 563–572.

86. R. Pan, B. Prabhakar, and K. Psounis, *CHOKe, a stateless active queue management scheme for approximating fair bandwidth allocation*, Proceedings of the IEEE INFOCOM (Tel Aviv, Israel), March 2000, pp. 942–951.

87. K. K. Ramakrishnan and R. Jain, *A binary feedback scheme for congestion avoidance in computer networks with a connectionless network layer*, Proceedings of ACM Sigcomm, 1988, pp. 303–313.

88. S. M. Ross, *Stochastic processes*, John Wiley & Sons, 1995.

89. N. Rouche, P. Habets, and M. Laloy, *Stability theory by Liapunov's direct method*, Springer-Verlag, 1977.

90. S. Shakkottai and R. Srikant, *Mean FDE models for Internet congestion control*, 2001, To appear in the *IEEE Transactions on Information Theory*. A shorter version appeared in the *Proceedings of IEEE Infocom*, 2002 under the title "How good are fluid models of Internet congestion control?".

91. P. Tinnakornsrisuphap and A. M. Makowski, *Limit behavior of ECN/RED gateways under a large number of TCP flows*, Proceedings of Infocom (San Francisco, CA), April 2003, pp. 873–883.

92. S. R. S. Varadhan, *Probability theory (Courant Lecture Notes, 7)*, American Mathematical Society, 2001.

93. G. Vinnicombe, *On the stability of end-to-end congestion control for the Internet*, 2001, University of Cambridge Technical Report CUED/F-INFENG/TR.398. Available at http://www.eng.cam.ac.uk/˜gv.

94. _____ , *On the stability of networks operating TCP-like congestion control*, Proceedings of the IFAC World Congress (Barcelona, Spain), 2002.

95. Z. Wang and F. Paganini, *Global stability with time-delay in network congestion control*, Proceedings of the IEEE Conference on Decision and Control, 2002.

96. J.T. Wen and M. Arcak, *A unifying passivity framework for network flow control*, Proceedings of IEEE Infocom, April 2003.

97. W. Whitt, *Stochastic process limits*, Springer, 2002.

98. Y. Xia, D. Harrison, S. Kalyanaraman, K. Ramachandran, and A. Venkatesan, *Accumulation-based congestion control*, 2002, RPI Technical Report.

99. H. Yaiche, R. R. Mazumdar, and C. Rosenberg, *A game-theoretic framework for bandwidth allocation and pricing in broadband networks*, IEEE/ACM Transactions on Networking **8** (2000), no. 5, 667–678.

Index

CPSIA information can be obtained at www.ICGtesting.com
Printed in the USA
LVOW070341240112

265289LV00007B/19/A